Science and Technology Concepts–Secondary™

Exploring the Properties of Matter

Student Guide

National Science Resources Center

The National Science Resources Center (NSRC) is operated by the Smithsonian Institution to improve the teaching of science in the nation's schools. The NSRC disseminates information about exemplary teaching resources, develops curriculum materials, and conducts outreach programs of leadership development and technical assistance to help school districts implement inquiry-centered science programs.

Smithsonian Institution

The Smithsonian Institution was created by an act of Congress in 1846 "for the increase and diffusion of knowledge..." This independent federal establishment is the world's largest museum complex and is responsible for public and scholarly activities, exhibitions, and research projects nationwide and overseas. Among the objectives of the Smithsonian is the application of its unique resources to enhance elementary and secondary education.

STC Program™ Project Sponsors

National Science Foundation

Bristol-Meyers Squibb Foundation

Dow Chemical Company

DuPont Company

Hewlett-Packard Company

The Robert Wood Johnson Foundation

Carolina Biological Supply Company

Science and Technology Concepts–Secondary™

Exploring the Properties of Matter

Student Guide

The STC Program™

Published by Carolina Biological Supply Company
Burlington, North Carolina

NOTICE This material is based upon work supported by the National Science Foundation under Grant No. ESI-9618091. Any opinions, findings, and conclusions or recommendations expressed in this material are those of the authors and do not necessarily reflect views of the National Science Foundation or the Smithsonian Institution.

This project was supported, in part, by the **National Science Foundation**.
Opinions expressed are those of the authors and not necessarily those of the foundation.

 First Edition 2012
10 9 8 7 6 5

ISBN 978-1-4350-0683-6

Published by Carolina Biological Supply Company, 2700 York Road, Burlington, NC 27215.
Call toll free 1-800-334-5551.

1503

Science and Technology Concepts—Secondary™
Exploring the Properties of Matter

The following revision was based on the STC/MS™ module *Properties of Matter.*

Developer
Kitty Lou Smith

Scientific Reviewers
Jerry A. Bell
Senior Scientist, Education Division
American Chemical Society

Matthew L. Clarke
Photographic Materials Scientist
National Gallery of Art

Kathryn J. Hughes
Program Officer, Board on Chemical Sciences and Technology
The National Academies

Laurie Friedman
Visiting Lecturer, Department of Chemistry and Food Science
Framingham State University

Tina Masciangioli
Senior Program Officer, Board on Chemical Sciences and Technology
The National Academies

Illustrator
Susie Duckworth

Developer/Writer
Interactive Whiteboard Activities
Sandy Ledwell, Ed.D

Photo Research
Jane Martin
Devin Reese

National Science Resources Center Staff

Executive Director
Thomas Emrick

Program Specialist/Revision Manager
Elizabeth Klemick

Contractor, Curriculum Research and Development
Devin Reese

Publications Graphics Specialist
Heidi M. Kupke

Carolina Biological Supply Company Staff

Director of Product and Development
Cindy Morgan

Marketing Manager, STC-Secondary™
Jeff Frates

Curriculum Editors
Lauren Eggiman
Gary Metheny

Managing Editor, Curriculum Materials
Cindy Vines Bright

Publications Designers
Trey Foster
Charles Thacker
Weldon D. Washington II
Greg Willette

Science and Technology Concepts for Middle Schools™
Properties of Matter
Original Publication

Module Development Staff

Developer/Writer
David Marsland

Science Advisor
Michael John Tinnesand
Head, K-12 Science
American Chemical Society

Contributing Writers
Linda Harteker
Robert Taylor

Illustrator
Max-Karl Winkler

STC/MS™ Project Staff

Principal Investigators
Douglas Lapp, Executive Director, NSRC
Sally Goetz Shuler, Deputy Director, NSRC

Project Director
Kitty Lou Smith

Curriculum Developers
David Marsland
Henry Milne
Carol O'Donnell
Dane J. Toler

Illustration Coordinator
Max-Karl Winkler

Photo Editor
Janice Campion

Graphic Designer
Heidi M. Kupke

STC/MS™ Project Advisors

Judy Barille, Chemistry Teacher, Fairfax County, Virginia, Public Schools

Steve Christiansen, Science Instructional Specialist, Montgomery County, Maryland, Public Schools

John Collette, Director of Scientific Affairs (retired), DuPont Company

Cristine Creange, Biology Teacher, Fairfax County, Virginia, Public Schools

Robert DeHaan, Professor of Physiology, Emory University Medical School

Stan Doore, Meteorologist (retired), National Oceanic and Atmospheric Administration, National Weather Service

Ann Dorr, Earth Science Teacher (retired), Fairfax County, Virginia, Public Schools; Board Member, Minerals Information Institute

Yvonne Forsberg, Physiologist, Howard Hughes Medical Center

John Gastineau, Physics Consultant, Vernier Corporation

Patricia Hagan, Science Project Specialist, Montgomery County, Maryland, Public Schools

Alfred Hall, Staff Associate, Eisenhower Regional Consortium at Appalachian Educational Laboratory

Connie Hames, Geology Teacher, Stafford County, Virginia, Public Schools

Jayne Hart, Professor of Biology, George Mason University

Michelle Kipke, Director, Forum on Adolescence, Institute of Medicine

John Layman, Professor Emeritus of Physics, University of Maryland

Thomas Liao, Professor of Engineering, State University of New York at Stony Brook

Ian MacGregor, Senior Science Advisor, Geoscience Education, National Science Foundation

Ed Mathews, Physical Science Teacher, Fairfax County, Virginia, Public Schools

Ted Maxwell, Geomorphologist, National Air and Space Museum, Smithsonian Institution

Tom O'Haver, Professor of Chemistry/Science Education, University of Maryland

Robert Ridky, Professor of Geology, University of Maryland

Mary Alice Robinson, Science Teacher, Stafford County, Virginia, Public Schools

Bob Ryan, Chief Meteorologist, WRC Channel 4, Washington, D.C.

Michael John Tinnesand, Head, K-12 Science, American Chemical Society

Grant Woodwell, Professor of Geology, Mary Washington College

Thomas Wright, Geologist, National Museum of Natural History, Smithsonian Institution; U.S. Geological Survey (emeritus)

Acknowledgments

The National Science Resources Center gratefully acknowledges the following individuals and school systems for their assistance with the national field-testing of *Properties of Matter:*

Delaware

Site Coordinator
Suzanne Curry, Director of Curriculum
Office of Climate and Standards
Wilmington High School

Site Coordinator
Mary Anne Wells
Mathematics and Science Resource Center
University of Delaware, Wilmington

Jillann Hounsell, Teacher
Henry B. du Pont Middle School, Hockessin

Thomas Hounsell, Teacher
Henry B. du Pont Middle School, Hockessin

Martin J. Cresci, Teacher
Henry C. Conrad Middle School, Wilmington

Maryland

Secondary Science Coordinator
Montgomery County Public Schools
Gerry Consuegra

Site Coordinator
Patricia Hagan, Science Project Specialist
Montgomery County Public Schools

Vince Parada, Teacher
Rosa M. Parks Middle School, Olney

Pam Fountain, Teacher
Tilden Middle School, Rockville

Oregon

Site Coordinator
Angie Ruzicka
4J Schools—Eugene School District

Courtney Abbott, Teacher
Kelly Middle School, Eugene

Julie Hohenemser, Teacher
Cal Young Middle School, Eugene

Judy Francis, Teacher
Roosevelt Middle School, Eugene

Pennsylvania

Site Coordinator
James L. Smoyer
Boyce Middle School, Pittsburgh

Sherri Petrella, Teacher
David E. Williams Middle School, Coraopolis

Jennifer Robinson
David E. Williams Middle School, Coraopolis

Jean M. Austin
Fort Couch Middle School, Upper St. Clair

Tennessee

Site Coordinator
Jimmie Lou Lee
Center for Excellence for Research and Policy on Basic Skills
Tennessee State University, Nashville

Janet Zanetis, Teacher
Ellis Middle School, Hendersonville

Judy W. Laney, Teacher
Grassland Middle School, Franklin

Gary Mullican, Teacher
Central Middle School, Murfreesboro

Ann Orman, Teacher
West End Middle School, Nashville

The NSRC appreciates the contribution of its STC/MS project evaluation consultants—

Program Evaluation Research Group (PERG), Lesley College

Sabra Lee
Researcher, PERG

George Hein
Director (retired), PERG

Center for the Study of Testing, Evaluation, and Education Policy (CSTEEP), Boston College

Joseph Pedulla
Director, CSTEEP

Maryellen Harmon
Director (retired), CSTEEP

Preface

Community leaders and state and local school officials across the country are recognizing the need to implement science education programs consistent with the National Science Education Standards to attain the important national goal of scientific literacy for all students in the 21st century. The Standards present a bold vision of science education. They identify what students at various levels should know and be able to do. They also emphasize the importance of transforming the science curriculum to enable students to engage actively in scientific inquiry as a way to develop conceptual understanding as well as problem-solving skills.

The development of effective standards-based, inquiry-centered curriculum materials is a key step in achieving scientific literacy. The National Science Resources Center (NSRC) has responded to this challenge through Science and Technology Concepts-Secondary™. Prior to the development of these materials, there were very few science curriculum resources for secondary students that embodied scientific inquiry and hands-on learning. With the publication of STC-Secondary™, schools will have a rich set of curriculum resources to fill this need.

Since its founding in 1985, the NSRC has made many significant contributions to the goal of achieving scientific literacy for all students. In addition to developing Science and Technology Concepts-Elementary™—an inquiry-centered science curriculum for grades K through 6—the NSRC has been active in disseminating information on science teaching resources, preparing school district leaders to spearhead science education reform, and providing technical assistance to school districts. These programs have had a significant impact on science education throughout the country. The transformation of science education is a challenging task that will continue to require the kind of strategic thinking and insistence on excellence that the NSRC has demonstrated in all of its curriculum development and outreach programs. The Smithsonian Institution, our sponsoring organization, takes great pride in the publication of this exciting new science program for secondary students.

Letter to the Students

Dear Student,

The National Science Resources Center's (NSRC) mission is to improve the learning and teaching of science for K-12 students. As an organization of the Smithsonian Institution, the NSRC is dedicated to the establishment of effective science programs for all students. To contribute to that goal, the NSRC has developed and published two comprehensive, research-based science curriculum programs: Science and Technology Concepts-Elementary™ and Science and Technology Concepts-Secondary™.

By using the STC-Secondary™ curriculum materials, we know that you will build an understanding of important concepts in life, earth, and physical sciences; learn critical-thinking skills; and develop positive attitudes toward science and technology. The National Science Education Standards state that all secondary students "...should be provided opportunities to engage in full and partial inquiries.... With an appropriate curriculum and adequate instruction, ... students can develop the skills of investigation and the understanding that scientific inquiry is guided by knowledge, observations, ideas, and questions."

STC-Secondary also addresses the national technology standards published by the International Technology Education Association. Informed by research and guided by standards, the design of the STC-Secondary units addresses four critical goals:

- Use of effective student and teacher assessment strategies to improve learning and teaching
- Integration of literacy into the learning of science by giving students the lens of language to focus and clarify their thinking and activities
- Enhanced learning using new technologies to help students visualize processes and relationships that are normally invisible or difficult to understand
- Incorporation of strategies to actively engage parents to support the learning process

We hope that by using the STC-Secondary curriculum you will expand your interest, curiosity, and understanding about the world around you. We welcome comments from students and teachers about their experiences with the STC-Secondary program materials.

Thomas Emrick
Executive Director
National Science Resources Center

Navigating an STC–Secondary™ Student Guide

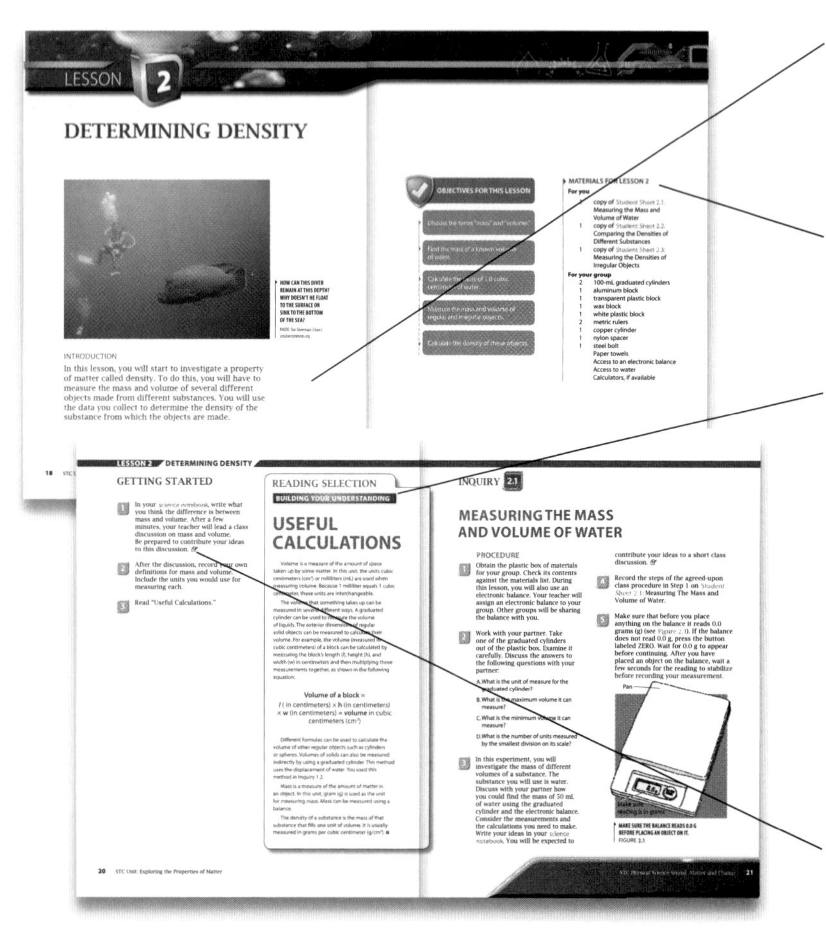

LESSON 2

DETERMINING DENSITY

HOW CAN THIS DIVER REMAIN AT THIS DEPTH? WHY DOESN'T HE FLOAT TO THE SURFACE OR SINK TO THE BOTTOM OF THE SEA?

INTRODUCTION

In this lesson, you will start to investigate a property of matter called density. To do this, you will have to measure the mass and volume of several different objects made from different substances. You will use the data you collect to determine the density of the substance from which the objects are made.

OBJECTIVES FOR THIS LESSON

- Discuss the terms "mass" and "volume."
- Find the mass of a known volume of water.
- Calculate the mass of 1.0 cubic centimeter of water.
- Measure the mass and volume of regular and irregular objects.
- Calculate the density of these objects.

MATERIALS FOR LESSON 2

For you

- 1 copy of Student Sheet 2.1: Measuring the Mass and Volume of Water
- 1 copy of Student Sheet 2.2: Comparing the Densities of Different Substances
- 1 copy of Student Sheet 2.3: Measuring the Densities of Irregular Objects

For your group

- 2 100-mL graduated cylinders
- 1 aluminum block
- 1 transparent plastic block
- 1 wax block
- 1 white plastic block
- 2 metric rulers
- 1 copper cylinder
- 1 nylon spacer
- 1 steel bolt
- Paper towels
- Access to an electronic balance
- Access to water
- Calculators, if available

18

LESSON 2 DETERMINING DENSITY

GETTING STARTED

1. In your science notebook, write what you think the difference is between mass and volume. After a few minutes, your teacher will lead a class discussion on mass and volume. Be prepared to contribute your ideas to this discussion.
2. After the discussion, record your own definitions for mass and volume. Include the units you would use for measuring each.
3. Read "Useful Calculations."

READING SELECTION

BUILDING YOUR UNDERSTANDING

USEFUL CALCULATIONS

Volume is a measure of the amount of space taken up by some matter. In this unit, the units cubic centimeters (cm^3) or milliliters (mL) are used when measuring volume. Because 1 milliliter equals 1 cubic centimeter, these units are interchangeable.

The volume that something takes up can be measured in several different ways. A graduated cylinder can be used to measure the volume of liquids. The exterior dimensions of regular solid objects can be measured to calculate their volume. For example, the volume (measured in cubic centimeters) of a block can be calculated by measuring the block's length (l), height (h), and width (w) in centimeters and then multiplying those measurements together, as shown in the following equation:

Volume of a block = l (in centimeters) × **h** (in centimeters) × **w** (in centimeters) = **volume** in cubic centimeters (cm^3)

Different formulas can be used to calculate the volume of other regular objects such as cylinders or spheres. Volumes of solids can also be measured indirectly by using a graduated cylinder. This method uses the displacement of water. You used this method in Inquiry 1.2.

Mass is a measure of the amount of matter in an object. In this unit, gram (g) is used as the unit for measuring mass. Mass can be measured using a balance.

The density of a substance is the mass of that substance that fills one unit of volume. It is usually measured in grams per cubic centimeter (g/cm^3). ■

20 STC Unit: Exploring the Properties of Matter

INQUIRY 2.1

MEASURING THE MASS AND VOLUME OF WATER

PROCEDURE

1. Obtain the plastic box of materials for your group. Check its contents against the materials list. During this lesson, you will also use an electronic balance. Your teacher will assign an electronic balance to your group. Other groups will be sharing the balance with you.
2. Work with your partner. Take one of the graduated cylinders out of the plastic box. Examine it carefully. Discuss the answers to the following questions with your partner:
   A. What is the unit of measure for the graduated cylinder?
   B. What is the maximum volume it can measure?
   C. What is the minimum volume it can measure?
   D. What is the number of units measured by the smallest division on its scale?
3. In this experiment, you will investigate the mass of different volumes of a substance. The substance you will use is water. Discuss with your partner how you could find the mass of 50 mL of water using the graduated cylinder and the electronic balance. Consider the measurements and the calculations you need to make. Write your ideas in your science notebook. You will be expected to contribute your ideas to a short class discussion.
4. Record the steps of the agreed-upon class procedure in Step 1 on Student Sheet 2.1: Measuring The Mass and Volume of Water.
5. Make sure that before you place anything on the balance it reads 0.0 grams (g) (see Figure 2.1). If the balance does not read 0.0 g, press the button labeled ZERO. Wait for 0.0 g to appear before continuing. After you have placed an object on the balance, wait a few seconds for the reading to stabilize before recording your measurement.

Pan

MAKE SURE THE BALANCE READS 0.0 G BEFORE PLACING AN OBJECT ON IT.
FIGURE 2.1

21

INTRODUCTION

This short paragraph helps get you interested about the upcoming inquiries.

MATERIALS

This helps you get organized and prepare for your inquiries.

READING SELECTION: BUILDING YOUR UNDERSTANDING

These reading selections are part of the lesson, and give you information about the topic or concept you are exploring.

NOTEBOOK ICON

During the course of an inquiry, you'll record data in different ways. This icon lets you know to record in your science notebook. Student sheets are called out when you're to write there. You may go back and forth between your notebook and a student sheet. Watch carefully for the icon throughout the procedure.

SAFETY TIPS

Safety in the science classroom is very important. Tips throughout the student guide will help you to practice safe techniques while conducting investigations. It is very important to read and follow all safety tips.

PROCEDURE

This tells you what to do. Sometimes the steps are very specific, and sometimes they guide you to come up with your own investigation and ways to record data.

REFLECTING ON WHAT YOU'VE DONE

These questions help you think about what you've learned during the lesson's inquiries, apply them to different situations, and generate new questions. Often you'll discuss your ideas with the class.

READING SELECTION: EXTENDING YOUR KNOWLEDGE

These reading selections come after the lesson, and show new ways that the topic or concept you learned about during the lesson can be applied, often in real-world situations.

GLOSSARY

Here you can find scientific terms defined.

INDEX

Locate specific information within the student guide using the index.

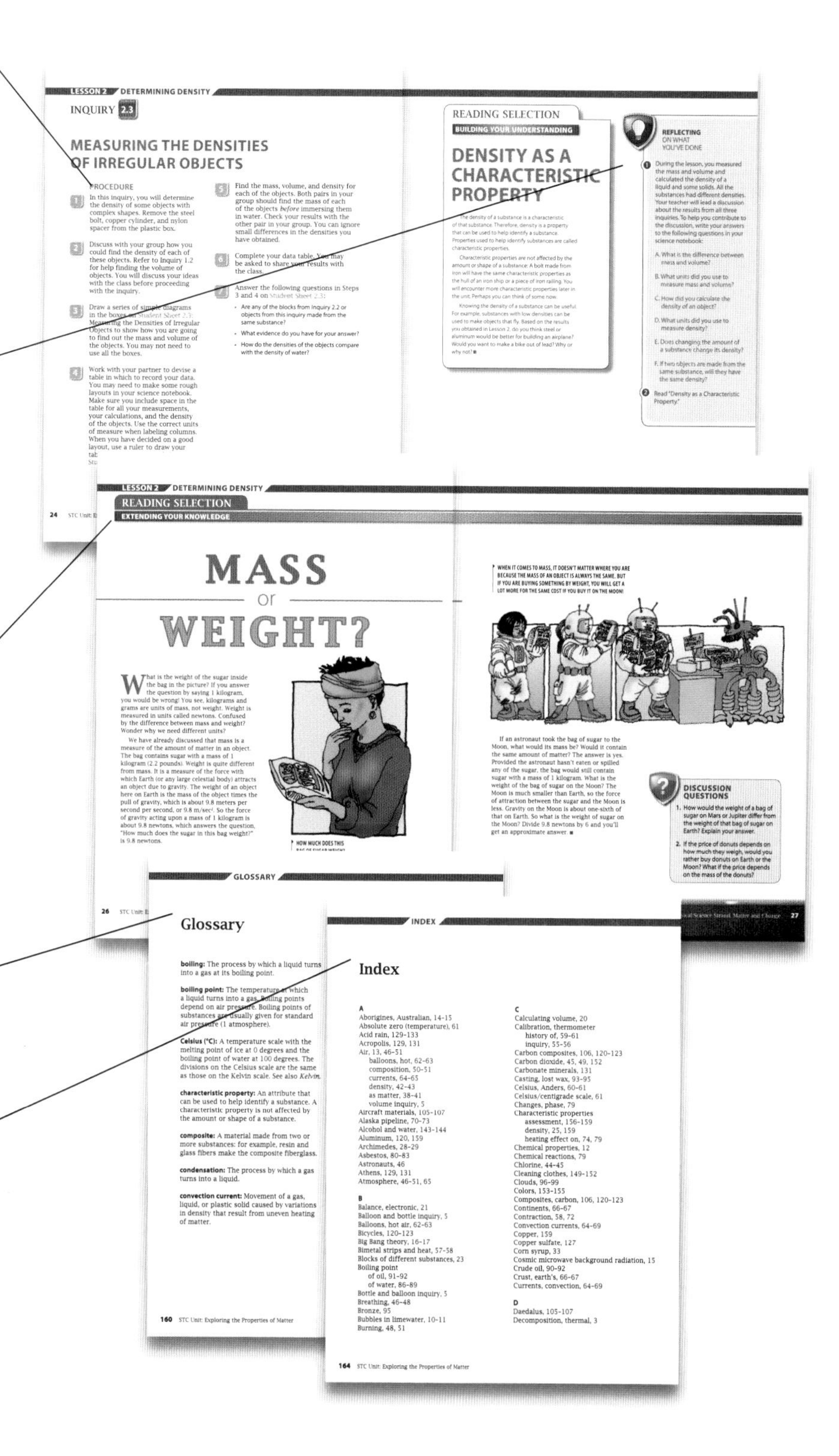

Contents

LESSON 1

OUR IDEAS ABOUT MATTER

MATTER IS EVERYWHERE ON EARTH. WHAT TYPE OF MATTER IS MOVING THESE TREES?

PHOTO: NOAA's National Weather Service Collection

INTRODUCTION

What is the meaning of the term "matter"? In this lesson, you will discuss the different meanings of this word and how it is used in science. You will also perform a circuit of eight inquiries on the properties of matter. These inquiries are designed to get you thinking about what matter is, what its properties are, and how it behaves. The observations you make and ideas you discuss in this lesson will play a key role in the inquiries that take place later in this unit.

OBJECTIVES FOR THIS LESSON

- Discuss your definitions of the term "matter."
- Observe some properties of matter.
- Use your own words and ideas to explain these properties.

MATERIALS FOR LESSON 1

For you

1 copy of Student Sheet 1: Our Ideas About Matter
1 pair of safety goggles

GETTING STARTED

1. Your teacher will place you in groups of four. What does your group think the word "matter" means? Write all the definitions you can think of on Student Sheet 1: Our Ideas About Matter.

2. After you have written down your definitions, your teacher will conduct a brainstorming session. Write the definition of matter that your class has agreed on.

3. In this lesson, you will complete a circuit of inquiries to investigate different properties of matter. Working with another student, you will observe one or more properties of matter at each station. Each pair of students will start at a different inquiry station. Your teacher will tell you at which station to begin.

4. Each inquiry has instructions you need to follow and questions you should try to answer. The procedure for each inquiry follows (see pages 5-11). It is also printed on a card placed at each station. When you make observations or think you can explain what you are observing, you should discuss these ideas with your partner. Remember: Exchanging ideas with others is a very important part of science.

5. In your own words, you will write your observations, explanations, and ideas for each activity on Student Sheet 1. When your teacher has assigned your group its first inquiry, find the inquiry number on your student sheet and place an asterisk (*) by it. This will help you to keep your observations in order.

6. When you have completed each inquiry, put the apparatus back as you found it at the beginning of the experiment.

7. You will have only 4–5 minutes at each station. When your teacher calls time, you must immediately go to the next experiment in the circuit. For example, if you are at Inquiry 1.3, move to 1.4, or if you are at Inquiry 1.8, move to 1.1.

SAFETY TIPS

Wear your safety goggles throughout the lesson.

The purple substance in Inquiry 1.4 can stain clothing and hands; use forceps when handling this substance.

If you are allergic to latex, notify your teacher.

Never taste a chemical or unknown substance in the science lab.

INQUIRY 1.1

THE BOTTLE AND THE BALLOON

PROCEDURE

1. Hold the bottle with the balloon attached to the top in the pot of hot water for 2 minutes (see Figure 1.1). Write a description of what happened when you placed the bottle in the pot of hot water.

2. Hold the bottle in the ice water for 1 minute. (If the ice has melted, you may need to add more.) What happened when you placed the bottle in the ice water?

3. Explain your observations in writing.

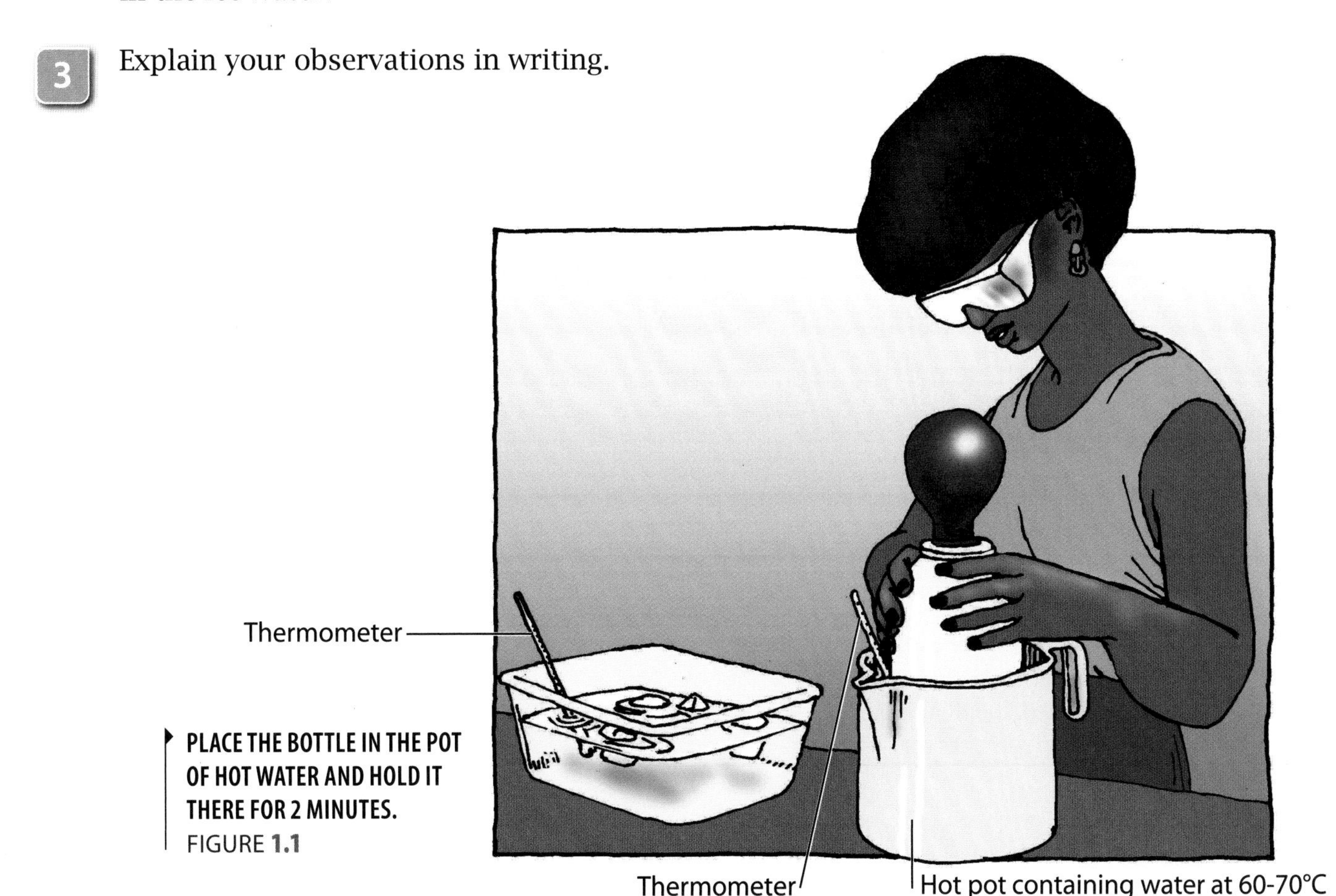

PLACE THE BOTTLE IN THE POT OF HOT WATER AND HOLD IT THERE FOR 2 MINUTES.
FIGURE **1.1**

INQUIRY 1.2

SIMILAR OBJECTS

PROCEDURE

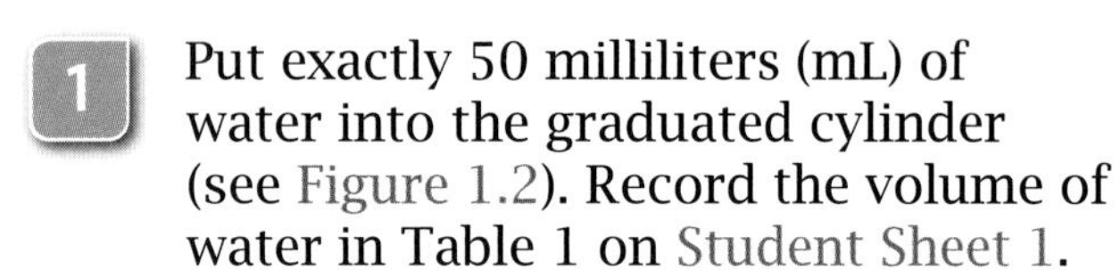

1. Put exactly 50 milliliters (mL) of water into the graduated cylinder (see Figure 1.2). Record the volume of water in Table 1 on Student Sheet 1.

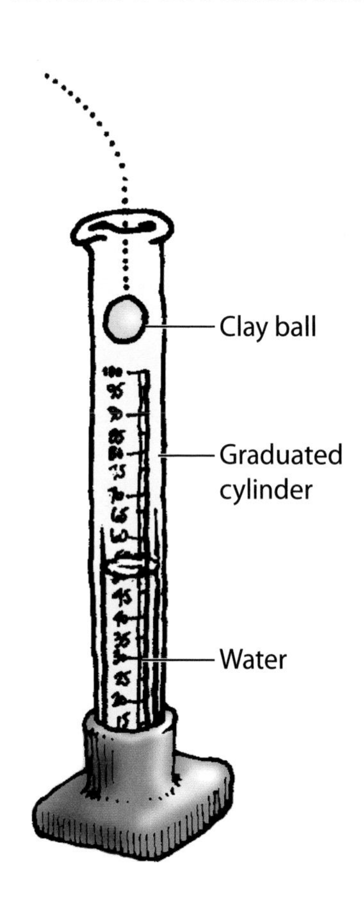

FILL EXACTLY HALF OF THE GRADUATED CYLINDER WITH WATER. MEASURE AND RECORD THE VOLUME OF WATER BEFORE ADDING THE BALL.
FIGURE **1.2**

2. Put the ball into the cylinder. Avoid splashing the water.

3. Record the volume of the water and the ball in Table 1.

4. Repeat the procedure using the rectangular block.

5. Use the data you have collected to calculate the volumes of the two objects. Record your results in Table 1.

6. How could you find out whether the two objects contain the same amount of matter? Record your ideas.

7. Return the apparatus to its original condition for the next group.

INQUIRY 1.3

TESTING SALT

PROCEDURE

1. Put 10 milliliters (mL) of water from the beaker labeled "Water" into the graduated cylinder and pour it into the test tube labeled "Water."

2. Place the test tube in the rack.

3. Repeat with 10 mL of oil, measuring the oil from the beaker marked "Oil" in the graduated cylinder and pouring it into the second test tube, labeled "Oil."

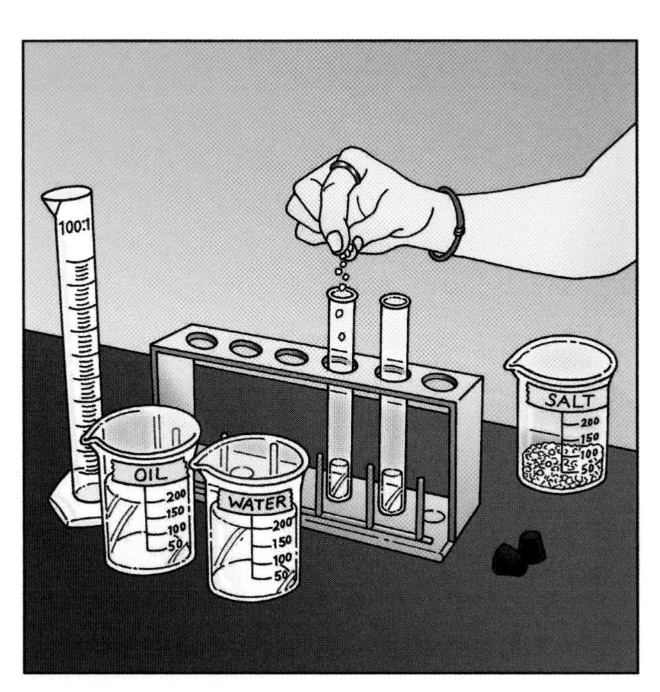

ADD EQUAL AMOUNTS OF SALT TO EQUAL AMOUNTS OF WATER AND OIL BEFORE TESTING ITS SOLUBILITY IN EACH LIQUID.
FIGURE **1.3**

4. Use the scoop to remove a small amount of salt from the beaker labeled "Salt" and add a pinch of salt to each test tube (see Figure 1.3).

5. Return the extra salt to the "Salt" beaker.

6. Observe each test tube carefully and compare the amount and appearance of the salt in each tube. Write your observations on Student Sheet 1.

7. Place a solid rubber stopper on each test tube. Shake each test tube vigorously for 2 minutes and observe. Be careful that you do not knock your test tube against your desk or table. How does the amount of salt in each of the two test tubes compare before and after shaking? How does the amount of salt in oil compare with the amount in water? Record your answers to these questions.

8. Shake the tubes again for 2 minutes and compare the amount of salt in each tube. Record your observations.

9. Restore the apparatus to its original condition for the next group.

INQUIRY 1.4

DESCRIBING MATTER

PROCEDURE

1. Use a loupe to examine substances A and B (see Figure 1.4). Write a description of them in as much detail as possible.

2. Do you think either A or B is a pure substance? Write a justification of your answer.

HOLD THE LOUPE CLOSE TO YOUR EYE AND TO THE SUBSTANCE YOU ARE EXAMINING.
FIGURE 1.4

INQUIRY 1.5

ADDING WATER

PROCEDURE

1. Place a clean, dry petri dish on the table in front of you.
2. Use the lab scoop to place a few grains of substance A on one side of the dish.
3. Use forceps to place a crystal of substance B on the other side of the dish.
4. Use the loupe to examine each substance. Draw a picture of what you see in Table 2 on Student Sheet 1.
5. Use the pipette to slowly add 20 drops of water to each substance.
6. Look at the substances again using the loupe. Make a drawing of each substance in Table 2.
7. Describe what happened to each substance when water was added to it. How did the two substances behave differently after water was added? Record what you observed.
8. Place the used petri dish in the plastic box provided.

INQUIRY 1.6

MIXING LIQUIDS

PROCEDURE

1. Look at the contents of the bottle.
2. Shake the bottle two times. Allow it to stand for 2 minutes.
3. Write a description of what you observe.
4. What do you know about the two substances in the bottle? Using your observations, write down everything you know about the two substances.

INQUIRY 1.7

FLOATING AND SINKING

PROCEDURE

1. Place the flattened pan and the regular pan into the tank of water.
2. What did you observe about each pan after it was placed in the water? Record your observations.
3. Both pans are made from the same substance and have the same mass. Why do the pans behave differently? Record your answer.
4. Remove both pans from the water and place them on the desk for the next group.

INQUIRY 1.8

BUBBLES IN THE LIMEWATER

PROCEDURE

1. Pour 10 milliliters (mL) of water from the beaker labeled "Water" into the graduated cylinder and add it to the test tube labeled "Water." Place it in the test tube rack.
2. Pour 10 mL of limewater into the graduated cylinder and then pour it into the second test tube labeled "Limewater."
3. Place the second test tube in the rack.
4. How do the liquids in the two test tubes compare in appearance? Record your observations.
5. Place one straw in the test tube with water and blow through it for 2 minutes.
6. Observe the appearance of the water in the test tube after you have blown into it.

7. Repeat with the second straw in the limewater test tube.

8. How do the two liquids compare in appearance? Record your answer.

9. Continue to blow through the straw into each liquid for another minute. Record any changes you observe. What do you think was added to the liquids when you exhaled through the straw?

10. What could account for the difference in the appearance of the two liquids? Write a possible explanation.

11. When you finish, empty the test tubes, throw out the straws, and prepare the equipment for the next group.

REFLECTING ON WHAT YOU'VE DONE

1. With your original group, record your ideas about each experiment on the sheet of paper assigned by your teacher.
2. After one minute, move to the next sheet.
3. When you have completed all the sheets, your teacher will tape them to the wall. Look at the observations made by your classmates.
4. Your teacher will ask different groups to read the comments on the sheets. How frequently did your group agree with the comments made by your classmates?
5. Later in the unit, you will do a more thorough investigation of the concepts introduced in this lesson. As you look at your original comments on these sheets, you may want to change or add to some of your original ideas.

READING SELECTION

EXTENDING YOUR KNOWLEDGE

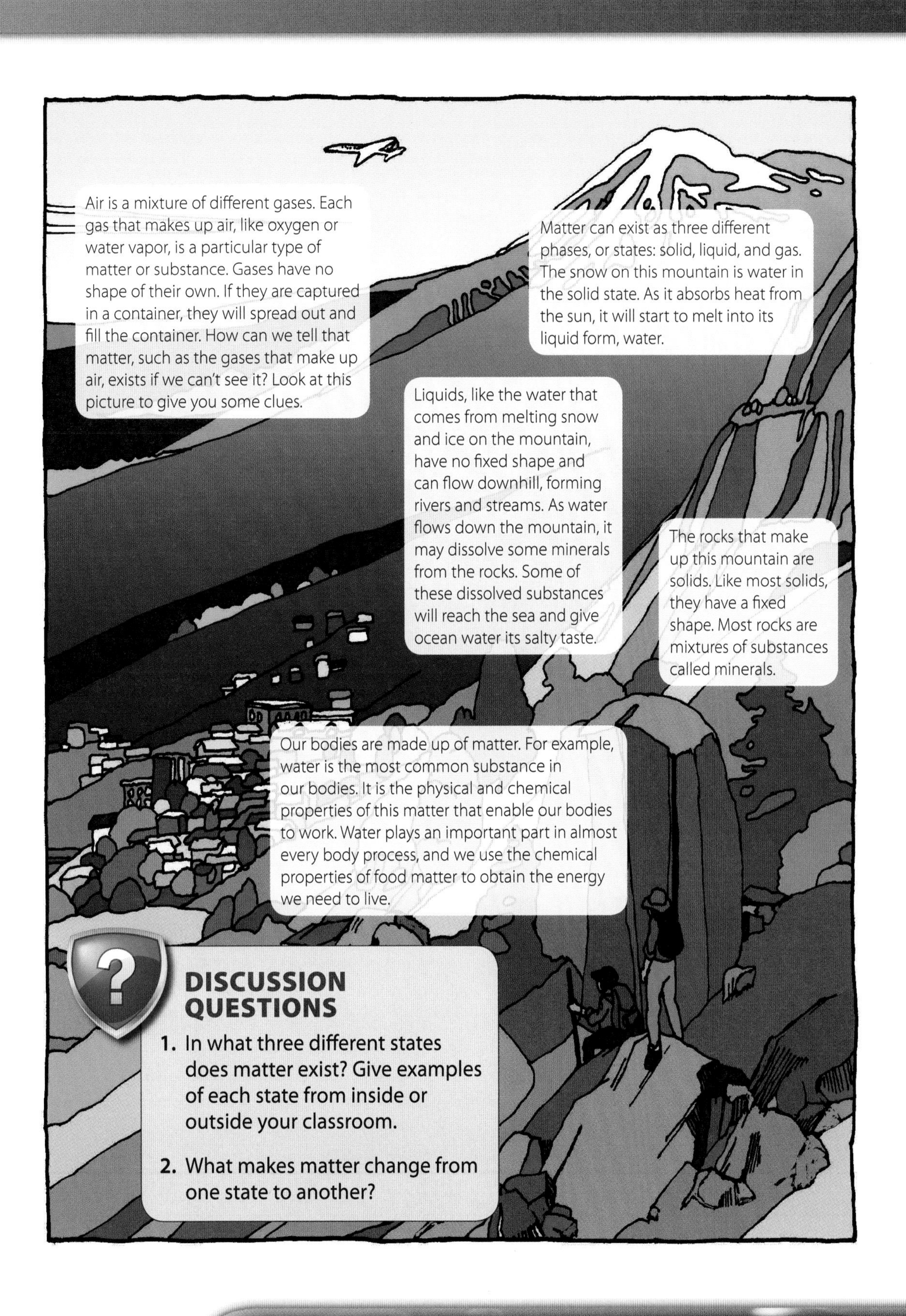

DISCUSSION QUESTIONS

1. In what three different states does matter exist? Give examples of each state from inside or outside your classroom.
2. What makes matter change from one state to another?

READING SELECTION

EXTENDING YOUR KNOWLEDGE

WHERE DID MATTER COME FROM?

Where did all the stuff in the universe—Earth, the sun, rocks, plants, animals, even you—come from? Was it made at a certain time? If so, how long has it been around? Has it always been there? People have asked these questions since the earliest times.

THESE AUSTRALIAN ABORIGINES ARE PAINTING IMAGES OF THEIR DREAMTIME STORY.

PHOTO: Jamie L. Fumo

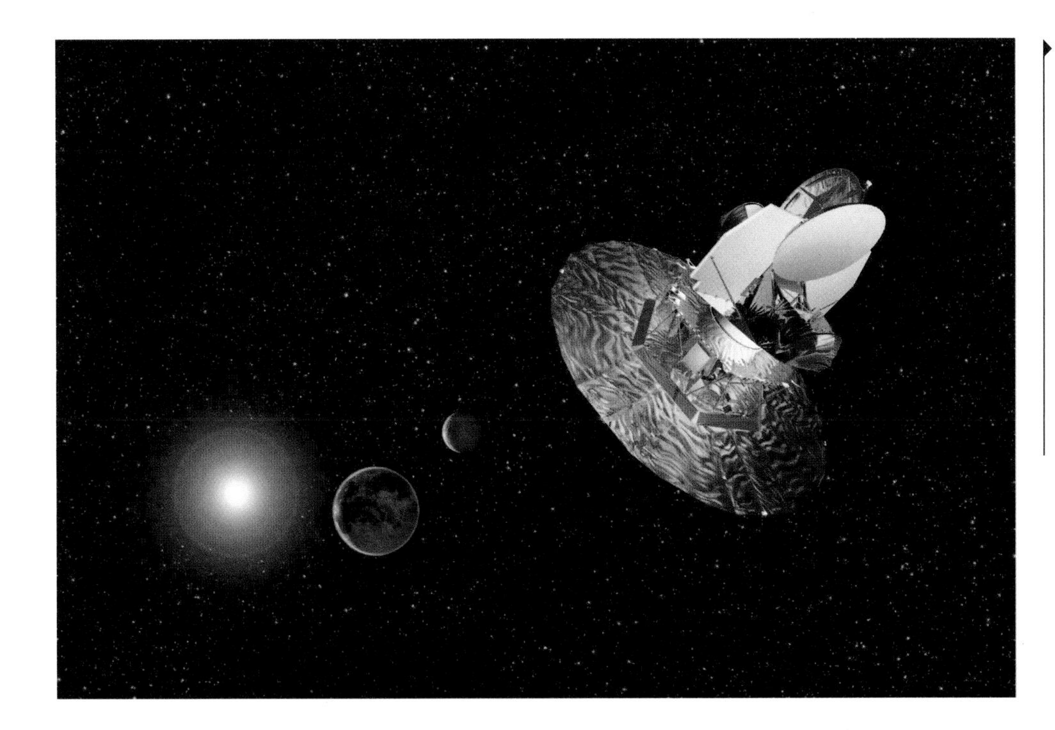

THE WILKINSON MICROWAVE ANISOTROPY PROBE (WMAP) SATELLITE WAS LAUNCHED IN 2001. ITS MISSION: TO FIND OUT MORE ABOUT THE COMPOSITION OF THE UNIVERSE BY STUDYING COSMIC MICROWAVE BACKGROUND RADIATION (THE HEAT LEFT OVER FROM THE BIG BANG).

PHOTO: NASA/WMAP Science Team

Most cultures have stories of how the universe was created. For instance, Australian Aborigines tell a story about the sun, moon, and stars sleeping beneath the ground. Their ancestors also slept there. One day the ancestors woke up and came to the surface. The Aborigines call this the Dreamtime. During the Dreamtime, the ancestors walked the earth as animals such as kangaroos, lizards, and wombats. Out of beings that were half animal and half plant, the ancestors made people. They then went back to sleep. Some went underground, but some became objects such as trees and rocks. The Dreamtime is an important part of Australian Aboriginal culture.

Scientists have also tried to answer the question of how the universe began. When scientists try to answer questions, they sometimes make observations of what happens and collect data. Scientists use their observations and data to try to explain the phenomena they are studying. One important part of science is that different scientists can make the same observations and collect the same data independently when they are studying the same phenomena. When the observations of many scientists agree over time, confirming one another, they may lead to the development of an accepted explanation or "theory." A theory allows scientists to make predictions and as new knowledge is gained, theories are tested and retested. Sometimes the theories don't stand up to the new information. These theories are then replaced with new theories.

READING SELECTION

EXTENDING YOUR KNOWLEDGE

Over the years, as scientists have gained new knowledge about the universe, new theories have replaced old theories. Currently, most scientists think the universe started with the "Big Bang." The Big Bang theory suggests that all the matter and energy in the universe exploded out from one point. As the explosion occurred, energy and matter spread outward and formed the universe. The matter from the Big Bang formed clouds of gas. As these gases cooled and condensed, stars, galaxies, and eventually planets and other "structures" that make up the universe were formed.

Even now, billions of years after the Big Bang, the universe is still spreading out. By looking at light from distant stars and galaxies, scientists can observe and measure this expansion. Using special apparatus, they can also detect some of the background glow of invisible energy left over from the Big Bang.

Space telescopes, such as the Hubble telescope and X-ray telescopes in orbit around Earth, are constantly making new and exciting observations. Ideas about the formation of the universe and the Big Bang may change as these instruments are used to discover more about our evolving universe. ■

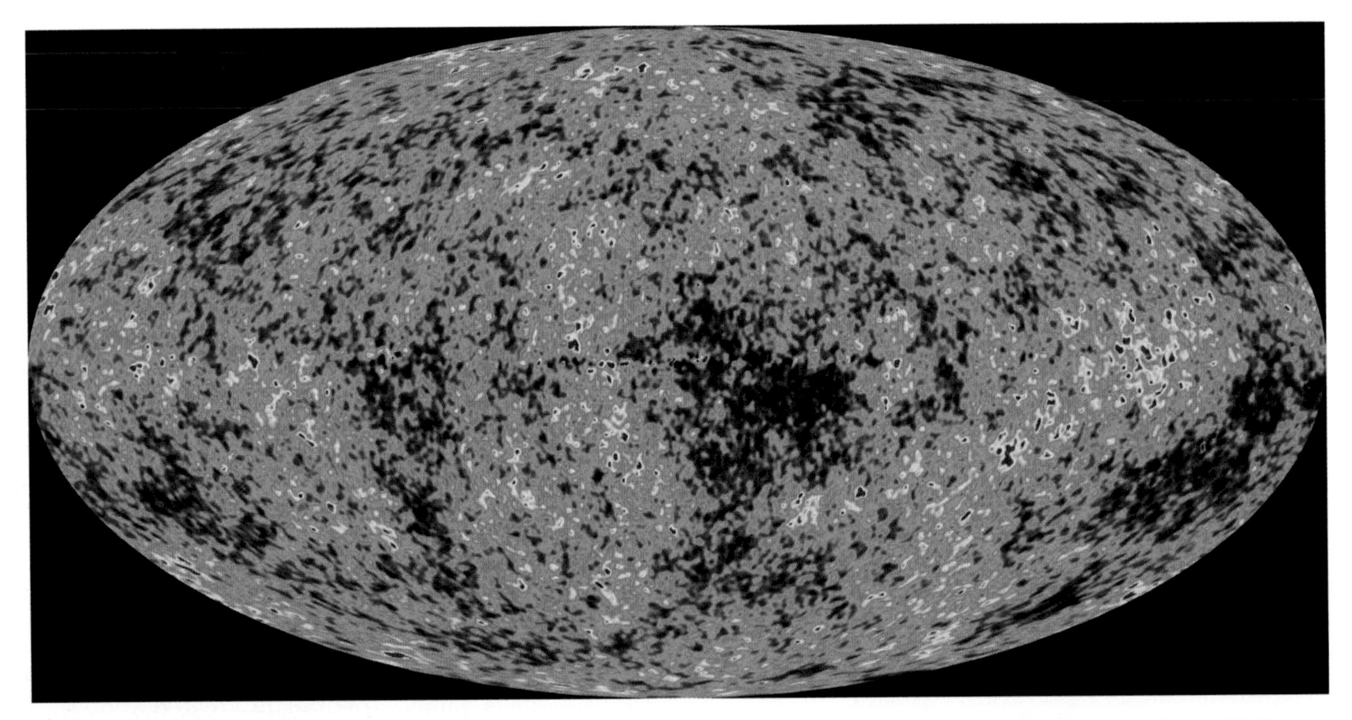

THIS MICROWAVE RADIATION MAP OF THE ENTIRE SKY WAS PRODUCED USING DATA COLLECTED BY WMAP. THE COLOR DIFFERENCES REPRESENT TEMPERATURE FLUCTUATIONS FROM RADIATION THAT WAS EMITTED 13.7 BILLION YEARS AGO, FOLLOWING THE BIG BANG. THIS IMAGE SHOWS THAT STRUCTURES SUCH AS GALAXIES WERE FORMING FROM THE TIME THE UNIVERSE BEGAN.

PHOTO: NASA/WMAP Science Team

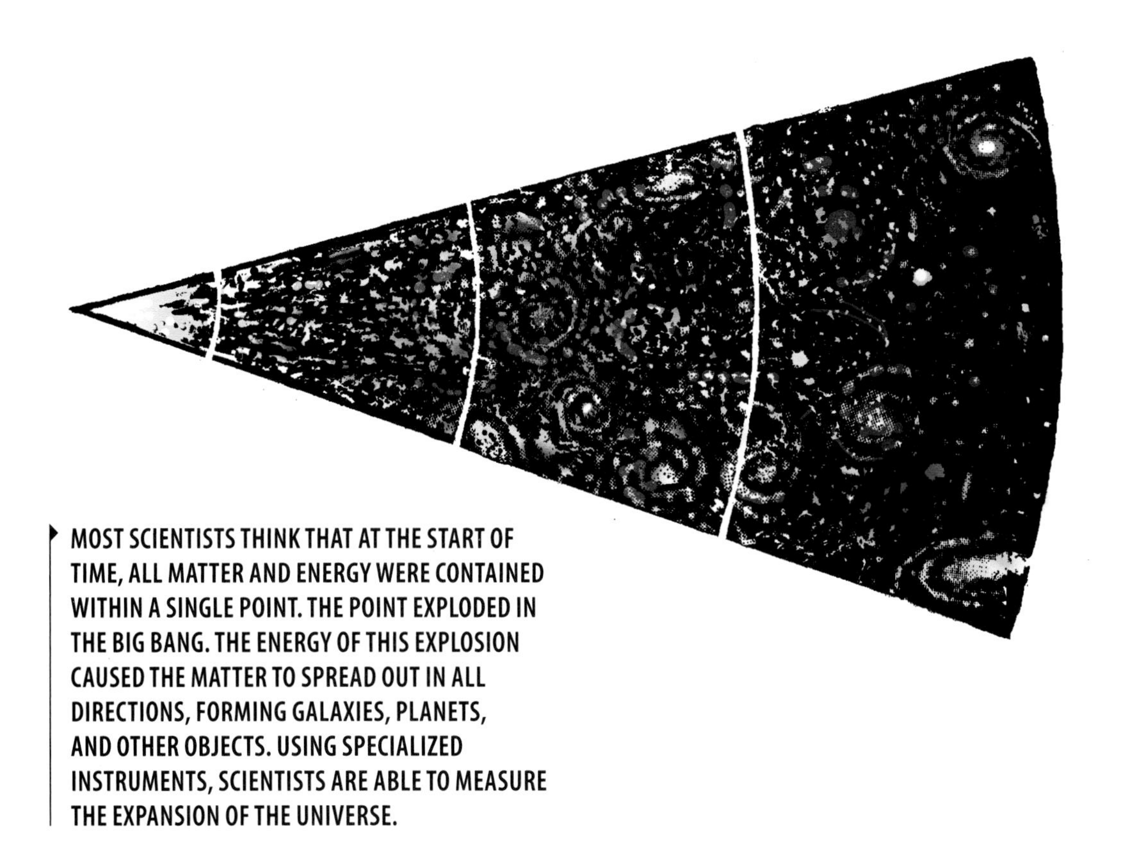

MOST SCIENTISTS THINK THAT AT THE START OF TIME, ALL MATTER AND ENERGY WERE CONTAINED WITHIN A SINGLE POINT. THE POINT EXPLODED IN THE BIG BANG. THE ENERGY OF THIS EXPLOSION CAUSED THE MATTER TO SPREAD OUT IN ALL DIRECTIONS, FORMING GALAXIES, PLANETS, AND OTHER OBJECTS. USING SPECIALIZED INSTRUMENTS, SCIENTISTS ARE ABLE TO MEASURE THE EXPANSION OF THE UNIVERSE.

DISCUSSION QUESTIONS

1. Use information from this reading selection, along with library and Internet resources, to list evidence for the Big Bang Theory.
2. In science, the term "theory" has a special meaning. Find a definition of this term, and give two examples of theories that are used in physical science.

LESSON 2

DETERMINING DENSITY

HOW CAN THIS DIVER REMAIN AT THIS DEPTH? WHY DOESN'T HE FLOAT TO THE SURFACE OR SINK TO THE BOTTOM OF THE SEA?

PHOTO: Tim Sheerman-Chase/ creativecommons.org

INTRODUCTION

In this lesson, you will start to investigate a property of matter called density. To do this, you will have to measure the mass and volume of several different objects made from different substances. You will use the data you collect to determine the density of the substance from which the objects are made.

OBJECTIVES FOR THIS LESSON

- Discuss the terms "mass" and "volume."
- Find the mass of a known volume of water.
- Calculate the mass of 1.0 cubic centimeter of water.
- Measure the mass and volume of regular and irregular objects.
- Calculate the density of these objects.

MATERIALS FOR LESSON 2

For you

1 copy of Student Sheet 2.1: Measuring the Mass and Volume of Water
1 copy of Student Sheet 2.2: Comparing the Densities of Different Substances
1 copy of Student Sheet 2.3: Measuring the Densities of Irregular Objects

For your group

2 100-mL graduated cylinders
1 aluminum block
1 transparent plastic block
1 wax block
1 white plastic block
2 metric rulers
1 copper cylinder
1 nylon spacer
1 steel bolt
Paper towels
Access to an electronic balance
Access to water
Calculators, if available

GETTING STARTED

1. In your science notebook, write what you think the difference is between mass and volume. After a few minutes, your teacher will lead a class discussion on mass and volume. Be prepared to contribute your ideas to this discussion.

2. After the discussion, record your own definitions for mass and volume. Include the units you would use for measuring each.

3. Read "Useful Calculations."

READING SELECTION

BUILDING YOUR UNDERSTANDING

USEFUL CALCULATIONS

Volume is a measure of the amount of space taken up by some matter. In this unit, the units cubic centimeters (cm^3) or milliliters (mL) are used when measuring volume. Because 1 milliliter equals 1 cubic centimeter, these units are interchangeable.

The volume that something takes up can be measured in several different ways. A graduated cylinder can be used to measure the volume of liquids. The exterior dimensions of regular solid objects can be measured to calculate their volume. For example, the volume (measured in cubic centimeters) of a block can be calculated by measuring the block's length (*l*), height (h), and width (w) in centimeters and then multiplying those measurements together, as shown in the following equation:

Volume of a block =
l (in centimeters) × **h** (in centimeters) × **w** (in centimeters) = **volume** in cubic centimeters (cm^3)

Different formulas can be used to calculate the volume of other regular objects such as cylinders or spheres. Volumes of solids can also be measured indirectly by using a graduated cylinder. This method uses the displacement of water. You used this method in Inquiry 1.2.

Mass is a measure of the amount of matter in an object. In this unit, gram (g) is used as the unit for measuring mass. Mass can be measured using a balance.

The density of a substance is the mass of that substance that fills one unit of volume. It is usually measured in grams per cubic centimeter (g/cm^3). ■

INQUIRY 2.1

MEASURING THE MASS AND VOLUME OF WATER

PROCEDURE

1. Obtain the plastic box of materials for your group. Check its contents against the materials list. During this lesson, you will also use an electronic balance. Your teacher will assign an electronic balance to your group. Other groups will be sharing the balance with you.

2. Work with your partner. Take one of the graduated cylinders out of the plastic box. Examine it carefully. Discuss the answers to the following questions with your partner:

   A. What is the unit of measure for the graduated cylinder?

   B. What is the maximum volume it can measure?

   C. What is the minimum volume it can measure?

   D. What is the number of units measured by the smallest division on its scale?

3. In this experiment, you will investigate the mass of different volumes of a substance. The substance you will use is water. Discuss with your partner how you could find the mass of 50 mL of water using the graduated cylinder and the electronic balance. Consider the measurements and the calculations you need to make. Write your ideas in your science notebook. You will be expected to contribute your ideas to a short class discussion.

4. Record the steps of the agreed-upon class procedure in Step 1 on Student Sheet 2.1: Measuring The Mass and Volume of Water.

5. Make sure that before you place anything on the balance it reads 0.0 grams (g) (see Figure 2.1). If the balance does not read 0.0 g, press the button labeled ZERO. Wait for 0.0 g to appear before continuing. After you have placed an object on the balance, wait a few seconds for the reading to stabilize before recording your measurement.

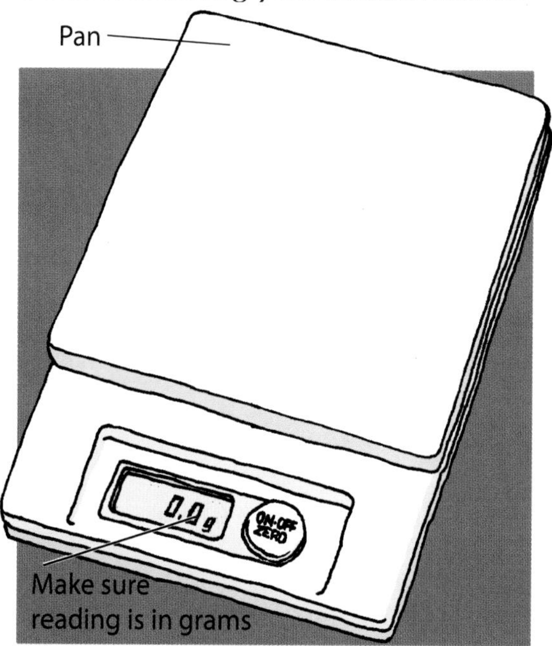

MAKE SURE THE BALANCE READS 0.0 G BEFORE PLACING AN OBJECT ON IT.
FIGURE **2.1**

Inquiry 2.1 continued

6. Look at Figure 2.2 to review how to accurately measure volume with a graduated cylinder.

7. Follow the class procedure to find the mass of 50 mL of water. Record your measurements in Table 1 on Student Sheet 2.1.

8. Complete the last column of Table 1. You can calculate the density of 1 cm^3 of water by dividing the mass of the water by the volume of the water.

9. Repeat the experiment using 25 mL of water. Remember, you already know the mass of the graduated cylinder.

10. Use your results to answer the questions in Steps 3-7 on Student Sheet 2.1.

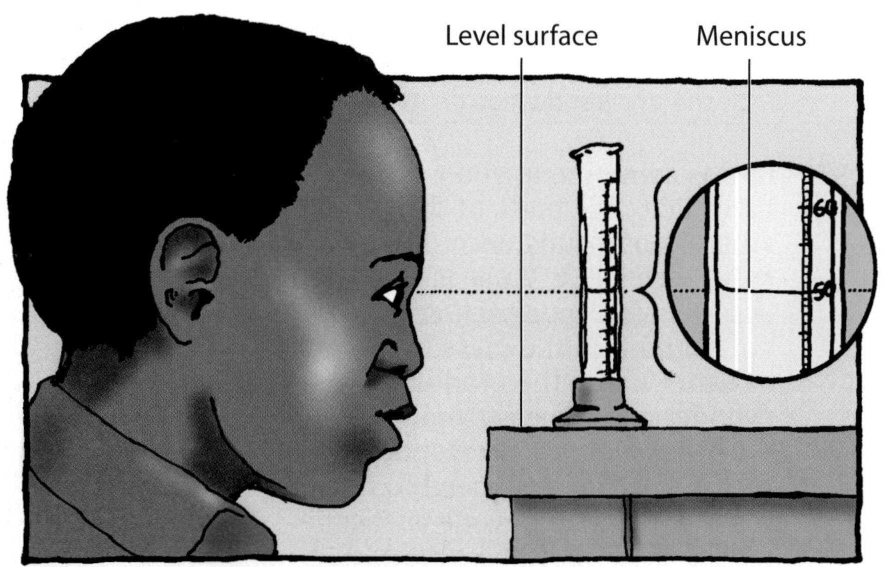

MAKE SURE THE GRADUATED CYLINDER IS ON A LEVEL SURFACE. WHEN YOU TAKE A READING, MAKE SURE YOUR EYE IS LEVEL WITH THE BOTTOM OF THE MENISCUS. THE MENISCUS IS THE CURVED UPPER SURFACE OF THE WATER IN THE CYLINDER.
FIGURE **2.2**

COMPARING THE DENSITIES OF DIFFERENT SUBSTANCES

PROCEDURE

1. Take the blocks of wax, transparent plastic, white plastic, and aluminum (the silver-colored metal) out of the plastic box.

2. Discuss the following questions with your group:

   A. Do you think all of these blocks have the same density?

   B. What evidence do you have to support your answer?

   C. What measurements will you need to make to test your hypothesis?

3. Work with your partner to determine the density of each of the blocks (see Figure 2.3). You will need to share the blocks with the other members of your group. Record your results in Table 1 on Student Sheet 2.2: Comparing the Densities of Different Substances.

4. Check your results and calculations against those of the other pair in your group. If your calculated densities do not match (within 0.1 g/cm^3), repeat your calculations.

5. Answer the following questions in Steps 2 and 3 on Student Sheet 2.2:
   - Are the densities of the different substances the same or different?
   - How could this information be used to identify the substance from which an object is made?

USE THE RULER TO MEASURE THE BLOCKS. CALCULATE THE VOLUME OF EACH BLOCK IN CUBIC CENTIMETERS. USE THE BALANCE TO MEASURE THE MASS OF THE BLOCKS.
FIGURE 2.3

INQUIRY 2.3

MEASURING THE DENSITIES OF IRREGULAR OBJECTS

PROCEDURE

1. In this inquiry, you will determine the density of some objects with complex shapes. Remove the steel bolt, copper cylinder, and nylon spacer from the plastic box.

2. Discuss with your group how you could find the density of each of these objects. Refer to Inquiry 1.2 for help finding the volume of objects. You will discuss your ideas with the class before proceeding with the inquiry.

3. Draw a series of simple diagrams in the boxes on Student Sheet 2.3: Measuring the Densities of Irregular Objects to show how you are going to find out the mass and volume of the objects. You may not need to use all the boxes.

4. Work with your partner to devise a table in which to record your data. You may need to make some rough layouts in your science notebook. Make sure you include space in the table for all your measurements, your calculations, and the density of the objects. Use the correct units of measure when labeling columns. When you have decided on a good layout, use a ruler to draw your table in the space provided on Student Sheet 2.3.

5. Find the mass, volume, and density for each of the objects. Both pairs in your group should find the mass of each of the objects *before* immersing them in water. Check your results with the other pair in your group. You can ignore small differences in the densities you have obtained.

6. Complete your data table. You may be asked to share your results with the class.

7. Answer the following questions in Steps 3 and 4 on Student Sheet 2.3:
   - Are any of the blocks from Inquiry 2.2 or objects from this inquiry made from the same substance?
   - What evidence do you have for your answer?
   - How do the densities of the objects compare with the density of water?

READING SELECTION

BUILDING YOUR UNDERSTANDING

DENSITY AS A CHARACTERISTIC PROPERTY

The density of a substance is a characteristic of that substance. Therefore, density is a property that can be used to help identify a substance. Properties used to help identify substances are called characteristic properties.

Characteristic properties are not affected by the amount or shape of a substance: A bolt made from iron will have the same characteristic properties as the hull of an iron ship or a piece of iron railing. You will encounter more characteristic properties later in the unit. Perhaps you can think of some now.

Knowing the density of a substance can be useful. For example, substances with low densities can be used to make objects that fly. Based on the results you obtained in Lesson 2, do you think steel or aluminum would be better for building an airplane? Would you want to make a bike out of lead? Why or why not? ■

REFLECTING
ON WHAT
YOU'VE DONE

1. During the lesson, you measured the mass and volume and calculated the density of a liquid and some solids. All the substances had different densities. Your teacher will lead a discussion about the results from all three inquiries. To help you contribute to the discussion, write your answers to the following questions in your science notebook:

   A. What is the difference between mass and volume?

   B. What units did you use to measure mass and volume?

   C. How did you calculate the density of an object?

   D. What units did you use to measure density?

   E. Does changing the amount of a substance change its density?

   F. If two objects are made from the same substance, will they have the same density?

2. Read "Density as a Characteristic Property."

READING SELECTION

EXTENDING YOUR KNOWLEDGE

MASS or WEIGHT?

What is the weight of the sugar inside the bag in the picture? If you answer the question by saying 1 kilogram, you would be wrong! You see, kilograms and grams are units of mass, not weight. Weight is measured in units called newtons. Confused by the difference between mass and weight? Wonder why we need different units?

We have already discussed that mass is a measure of the amount of matter in an object. The bag contains sugar with a mass of 1 kilogram (2.2 pounds). Weight is quite different from mass. It is a measure of the force with which Earth (or any large celestial body) attracts an object due to gravity. The weight of an object here on Earth is the mass of the object times the pull of gravity, which is about 9.8 meters per second per second, or 9.8 m/sec^2. So the force of gravity acting upon a mass of 1 kilogram is about 9.8 newtons, which answers the question, "How much does the sugar in this bag weigh?"

HOW MUCH DOES THIS BAG OF SUGAR WEIGH?

WHEN IT COMES TO MASS, IT DOESN'T MATTER WHERE YOU ARE BECAUSE THE MASS OF AN OBJECT IS ALWAYS THE SAME. BUT IF YOU ARE BUYING SOMETHING BY WEIGHT, YOU WILL GET A LOT MORE FOR THE SAME COST IF YOU BUY IT ON THE MOON!

If an astronaut took the bag of sugar to the Moon, what would its mass be? Would it contain the same amount of matter? The answer is yes. Provided the astronaut hasn't eaten or spilled any of the sugar, the bag would still contain sugar with a mass of 1 kilogram. What is the weight of the bag of sugar on the Moon? The Moon is much smaller than Earth, so the force of attraction between the sugar and the Moon is less. Gravity on the Moon is about one-sixth of that on Earth. So what is the weight of sugar on the Moon? Divide 9.8 newtons by 6 and you'll get an approximate answer. ■

DISCUSSION QUESTIONS

1. How would the weight of a bag of sugar on Mars or Jupiter differ from the weight of that bag of sugar on Earth? Explain your answer.
2. If the price of donuts depends on how much they weigh, would you rather buy donuts on Earth or the Moon? What if the price depends on the mass of the donuts?

READING SELECTION

EXTENDING YOUR KNOWLEDGE

ARCHIMEDES' Crowning Moment

Archimedes, one of the most famous mathematicians and scientists of ancient Greece, had a problem. The king had a new crown. It looked like pure gold. But the king was suspicious. How could he be sure that the jeweler hadn't cheated him by adding another, less valuable metal to the molten gold? The king asked Archimedes to find out whether the crown was made from pure gold.

Archimedes knew his reputation was on the line. He could have taken the problem down to the public marketplace, where he often went to discuss scientific questions with other scholars. But instead, he decided to relax in a bath. The tub was filled to the brim. Still concentrating on his problem, Archimedes immersed himself in the water.

Splash! Water spilled over the sides of the tub and onto the floor. He had made a real mess. But that mess triggered an idea—an idea that would help solve the king's dilemma.

"When I got into the tub," Archimedes reasoned, "my body displaced a lot of water. Now, there must be a relationship between my volume and the volume of water that my body displaced—because if I weren't so big, less water would have spilled on my floor."

This observation brought Archimedes back to the problem of the gold crown. What if he put it in water? How much water would it displace? And could he apply this observation to prove that the crown was made of pure gold?

"HMMM...THE VOLUME OF MY BODY EQUALS THE VOLUME OF THE WATER ON THE BATHROOM FLOOR."

ARCHIMEDES WAS AN EXPERT ON MASS, VOLUME, AND DENSITY.

PHOTO: Courtesy of the Smithsonian Institution Libraries, Dibner Library of the History of Science and Technology, Washington, D.C.

Archimedes knew about the importance of controls, so he began by finding a piece of gold and a piece of silver with exactly the same mass. He dropped the gold into a bowl filled to the brim with water and measured the volume of water that spilled out. Then he did the same thing with the piece of silver.

Although both metals had the same mass, the silver had a larger volume; therefore, it displaced more water than did the gold. That's because silver is less dense than gold.

Now it was time to check out the crown. Archimedes found a piece of pure gold that had the same mass as the crown. He placed the pure gold chunk and the crown in water one at a time, and measured the volume of water that spilled out.

The crown displaced more water than the piece of gold. Therefore, its density was less than pure gold. The king had been cheated! Although this was just one of Archimedes' many contributions to science, there's no doubt that it was his "crowning moment"! ■

DISCUSSION QUESTIONS

1. If Archimedes had only been able to find a piece of pure gold with a mass one-half that of the crown, how could he have determined whether the crown was pure gold?
2. If the crown had displaced less water than the piece of gold, what conclusions might Archimedes have drawn?

LESSON 3

DENSITY PREDICTIONS

HOW COULD YOUR KNOWLEDGE OF DENSITY BE USED TO HELP CLEAN UP THIS OIL SPILL?

PHOTO: U.S. Coast Guard photo by Public Affairs Specialist 3rd Class Adam Baylor

INTRODUCTION

Why is it useful to know about density? You have already discussed how density, because it is a characteristic property of matter, can be used as one way to help identify a substance. You can also use the density of an object or substance to predict how it may behave under different conditions. For example, have you ever done experiments that involved investigating whether objects float or sink in water? Apart from guessing, how can you tell whether an object will float or sink? Are there some measurements that can be used to predict floating and sinking? In this lesson, you will use the data you have already collected on density and relate it to floating and sinking. You will then use density to predict how solids and liquids behave in a density column.

OBJECTIVES FOR THIS LESSON

- Predict whether an object will float or sink based on how it feels.
- Use density to predict whether a substance will float or sink in water.
- Determine the density of different liquids.
- Build a density column.
- Use density to predict how solids will behave when they are placed in a density column.

MATERIALS FOR LESSON 3

For you

1 copy of Student Sheet 3.1: Using Density to Make Predictions
1 copy of Student Sheet 3: Homework for Lesson 3

For your group

2 100-mL graduated cylinders
1 250-mL beaker containing colored water
1 copper cylinder
1 nylon spacer
1 test tube brush
1 aluminum block
1 transparent plastic block
1 wax block
1 white plastic block
1 bottle of vegetable oil
1 bottle of corn syrup

For the class

1 container for collecting vegetable oil waste
1 container for collecting corn syrup and water waste
Paper towels
Detergent
Access to an electronic balance
Access to water

GETTING STARTED

1. Obtain the plastic box of materials for your group. Check its contents against the materials list. During this lesson, you will use an electronic balance. Your teacher will assign an electronic balance to your group. Other groups will share the balance with you.

2. Take the blocks of aluminum, wax, and white and transparent plastic out of the plastic box. As a group, predict whether each object will float or sink in water and explain how you reached your prediction.

3. Your teacher will list the predictions for each group and may ask you to explain your predictions to the class.

4. Your teacher will ask some of you to test your predictions. Use the results of these tests and the data you collected in Lesson 2 to fill in Table 1 on Student Sheet 3.1: Using Density to Make Predictions.

5. Is there a relationship between density and floating and sinking in water? If so, write a description of this relationship.

WHAT PROPERTY KEEPS THIS RAFT AFLOAT SO THE DOG CAN DRY OFF?

PHOTO: Brendan Adkins/creativecommons.org

INQUIRY 3.1

BUILDING A DENSITY COLUMN

PROCEDURE

1. In this inquiry, your group will work together using all the materials in the plastic box.

2. Look carefully at Table 2 on Student Sheet 3.1: Using Density to Make Predictions. You will need to determine the density of three liquids. You already have some of this information.

3. Spend a few minutes carefully reviewing the procedure you used in Inquiry 2.1 to determine the density of water.

4. Use the same procedure to find out the density of corn syrup and vegetable oil. Use 25 mL of each substance. Use a different graduated cylinder for each substance. The graduated cylinders may have different masses. Be sure to check the mass of each. Do not empty the cylinders; you will need both of the liquids later in this inquiry.

5. Use the data you collect to fill in Table 2.

6. Look carefully at the densities you have calculated. What do you predict will happen when you mix together the vegetable oil, corn syrup, and water? Record and explain your prediction, then fill in the "Prediction Cylinder" on Student Sheet 3.1.

7. Pour the 25 mL of vegetable oil into the cylinder containing the corn syrup. Add an additional 25 mL of colored water from the beaker. (The colored water has the same density as water.) Allow the contents of the cylinder to settle.

8. Fill in and label the "Observation Cylinder" on Student Sheet 3.1.

9. Do the liquids mix together (miscible) or form distinct layers (immiscible)? What is the relationship between the density of the liquid and its position in the graduated cylinder? Write your answers in Steps 7 and 8 on Student Sheet 3.1.

10. Use information you obtained in Inquiry 2.1 to predict what will happen when you drop the copper cylinder into your density column and when you drop the nylon spacer into the column. Discuss your ideas with your group.

Inquiry 3.1 continued

11. Drop the copper cylinder, followed by the nylon spacer, into the column. Observe what happens. Record your results in the "Observation Cylinder" on Student Sheet 3.1. Label each object and write down its density.

12. Carefully pour the vegetable oil into the container provided for this purpose; do the same with the syrup and water.

13. Using the test tube brush, thoroughly wash all the graduated cylinders and objects in a detergent solution.

14. Dry the objects with a paper towel. Stand the graduated cylinders upside down in a sink or on newspaper to allow them to drain.

15. If you have spilled any substances, wipe off your table.

REFLECTING ON WHAT YOU'VE DONE

1. Write a short paragraph in your science notebook explaining your observations. Make sure you include the words "density" and "miscible" or "immiscible" in your description. Be prepared to read your paragraph to the class.

2. Your teacher will show the bottle containing two liquids that you used in Inquiry 1.6. Use your knowledge of immiscible liquids and density to explain (in your notebook) the appearance and behavior of the liquids in the bottle.

3. Oil is less dense than water. With your group, discuss how this information can be applied to cleaning up a spill from an oil tanker.

Why Did the *Titanic* Float?

On April 10, 1912, the luxury liner *Titanic* left England for New York and sailed straight into the annals of history. Why is the name *Titanic* so well known? At that time, she was considered the safest ship ever built; some people even considered her unsinkable. The *Titanic* became famous when she struck an iceberg and sank on her first voyage. About 1,500 people drowned or froze to death in the ice-cold Atlantic water.

People often ask, "Why did the *Titanic* sink?" Perhaps a better question would be, "Why did the ship float?" She was, after all, made mainly from iron and steel. Her anchors alone weighed 28 metric tons. (That's almost 62,000 pounds!) Steel has a density about eight times that of water, so you would expect a ship made of steel to sink.

THE TITANIC NOW LIES UNDER ABOUT 3,800 METERS (ABOUT 12,467 FEET) OF WATER. IT WAS MADE MAINLY FROM STEEL, WHICH IS DENSER THAN WATER. HOW DID IT MANAGE TO FLOAT AT ALL?

PHOTO: Courtesy of NOAA and the Russian Academy of Sciences

READING SELECTION

EXTENDING YOUR KNOWLEDGE

The World.

"Circulation Books Open to All."

VOL. LII. NO. 18,501. NEW YORK, TUESDAY, APRIL 16, 1912.

GREAT TITANIC SINKS; MORE THAN 1.500 LOST; 866 WOMEN AND CHILDREN KNOWN TO BE SAVED; SCORES OF NOTABLES NOT ACCOUNTED FOR

THE LOST LINER, HER POSITION AND THAT OF OTHER SHIPS WHEN SHE HIT ICEBERG

The TITANIC
LENGTH - 882 FT
BEAM - 92 FT
DEPTH - 94 FT
DISPLACEMENT 45,000 TONS
VALUE (ESTIMATED) $10,000,000

WHITE LINES ON SIDE OF STEAMSHIP INDICATE LOCATION of BULKHEADS

600 MILES

White Star Official Admits the Greatest Disaster in Marine History—J. J. Astor Rumored Lost, but Bride Saved—Text of Olympic's Fateful Message—Partial List Is Received.

HOPE THAT MANY WILL BE FOUND ON WRECKAGE.

Virginian and Parisian Reach Scene Too Late—They Are Joined by Other Steamers, Which Find Only Debris—Capt. Smith Believed to Have Gone Down with Ship—The Saved Suffer Severely from Exposure, After Floating in the Lifeboats for Eight Hours.

LIST OF THE KNOWN SAVED

THIS NEWSPAPER ARTICLE REPORTED THE DISASTROUS MAIDEN VOYAGE OF THE TITANIC. WHY WAS THIS VOYAGE A DISASTER? WHAT ROLE DID DENSITY PLAY IN THE TRAGEDY?

PHOTO: Library of Congress, Prints & Photographs Division, LC-USZ62-116257

ICEBERGS FLOAT IN WATER. WHAT DOES THIS TELL US ABOUT THEIR DENSITY?

PHOTO: U.S. Coast Guard

However, if you were to look at a plan of the *Titanic*, you would discover that most of her volume was occupied by air. Air has a density of about one-thousandth that of water. Therefore, the average density of the ship was less than the density of water. That's why she floated.

Why did she sink? When the *Titanic* hit the iceberg, water rushed into the ship's hull and displaced the air. The average density of the water and the steel ship was greater than the density of water. The result of this change? The *Titanic* sank to the bottom of the Atlantic. ■

▶ PHOTO: U.S. Navy photo by Photographer's Mate 2nd Class Andrew M. Meyers

DISCUSSION QUESTIONS

1. Why do ships have hulls painted two different colors, as shown in the photograph at left? Hint: It has something to do with density.

2. In a disaster at sea, life vests (also called personal flotation devices, or PFDs) can save lives. Design and draw your own PFD, and label its parts. Explain the role of each part, and be sure to include the word "density" in your explanation.

LESSON 4

DO GASES HAVE DENSITY?

A DESTRUCTIVE TORNADO LIKE THIS ONE IS EVIDENCE THAT AIR IS MATTER.

PHOTO: OAR/ERL/National Severe Storms Laboratory (NSSL)

INTRODUCTION

Have you ever thought about air? Air is strange stuff. It's invisible, yet we know it exists. We can feel our own breath or see the effect of the wind. But is air matter? If it is, then it must have both mass and volume. In this lesson, you will find out whether air has mass and volume.

OBJECTIVES FOR THIS LESSON

- Find out whether air has volume.
- Design an experiment that can be used to find the mass of a sample of air.
- Try to measure the mass and volume of a sample of air and then calculate its density.
- Discuss the accuracy of the procedure.

MATERIALS FOR LESSON 4

For you

1 copy of Student Sheet 4.1: Finding the Density of Air

For your group

1 thick-walled plastic bottle
1 rubber washer
1 vacuum pump with vacuum stopper (rubber valve)
1 100-mL graduated cylinder
Access to water
Access to an electronic balance

GETTING STARTED

1. Your teacher will show you two pieces of apparatus. In the first one, the funnel goes into the test tube but is held firmly in place by a rubber stopper. The second test tube also holds a funnel, but is not sealed around the edge of the tube.

2. In your science notebook, describe what happens when colored water is poured into each funnel (see Figure 4.1). Try to explain why water behaves differently in each funnel.

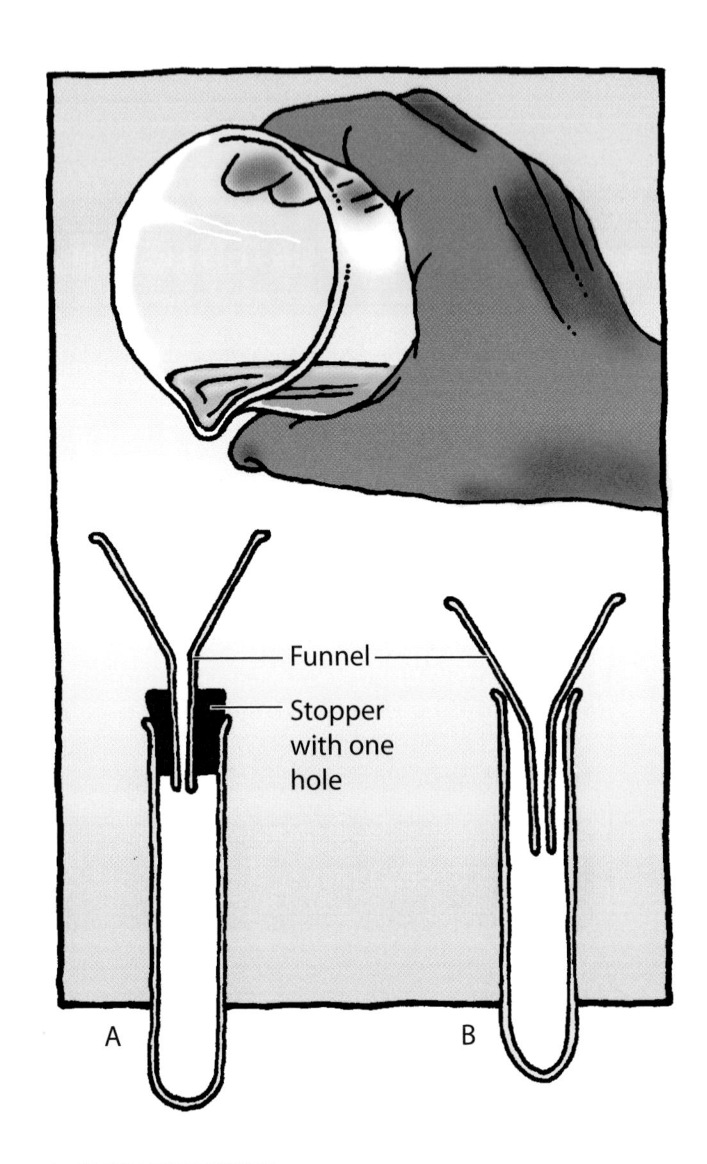

IN THIS EXPERIMENT, COLORED WATER IS POURED INTO BOTH FUNNELS.
FIGURE **4.1**

3. Your teacher will pass around the second piece of apparatus, which consists of two syringes connected by a tube.

4. Try to explain what you observe when the plunger of one syringe is pushed in. Draw what happens when the syringe is pushed in (see Figure 4.2).

5. Write a paragraph describing what these two experiments tell you about air.

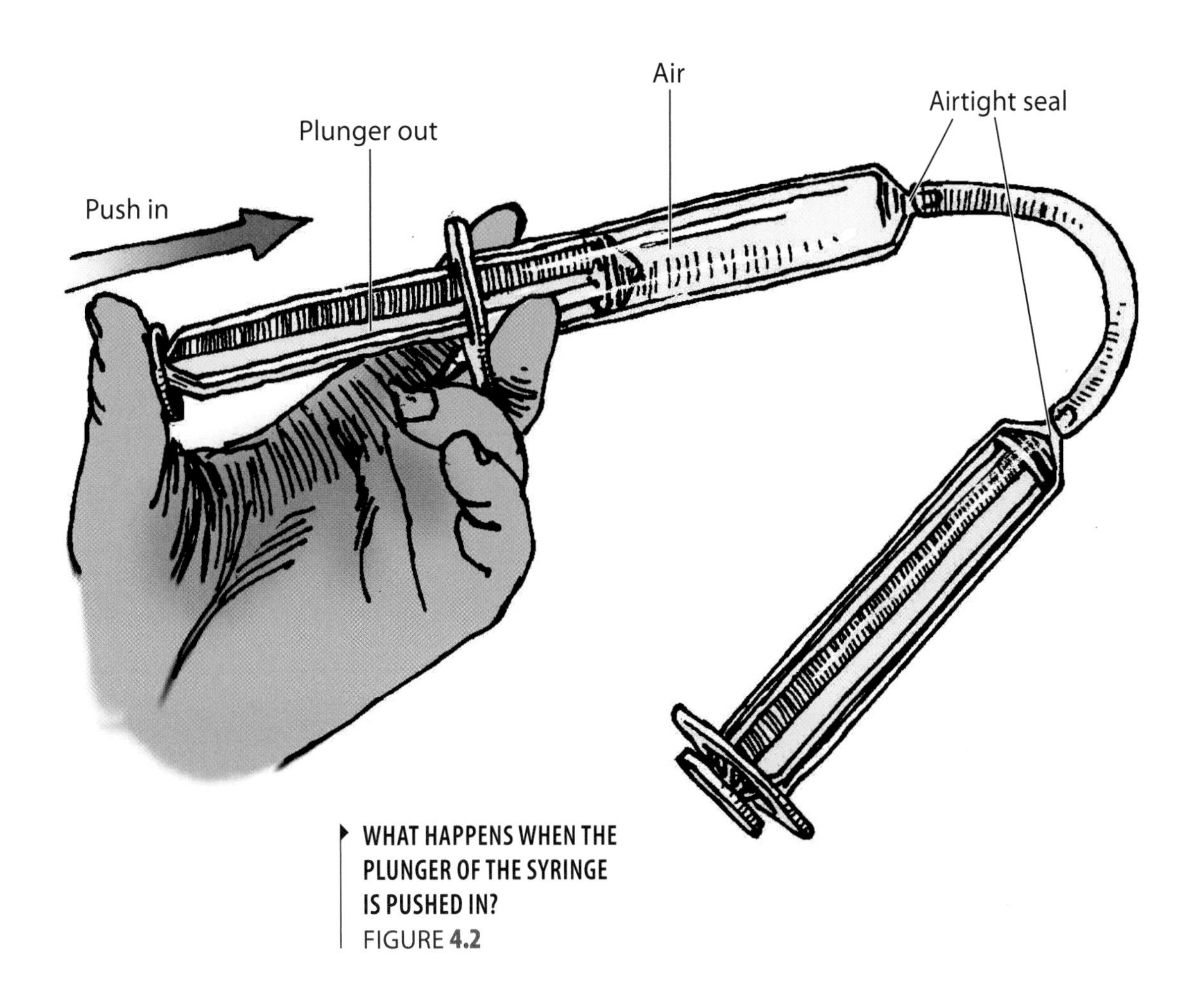

WHAT HAPPENS WHEN THE PLUNGER OF THE SYRINGE IS PUSHED IN?
FIGURE **4.2**

INQUIRY 4.1

FINDING THE DENSITY OF AIR

PROCEDURE

1. What will you need to measure in order to calculate the density of air? Examine the contents of your plastic box and try to think how you could use the apparatus and an electronic balance to find the density of some air (see Figure 4.3). Discuss your ideas with your group. Try to agree on a procedure that you think will work. Write your ideas in a short paragraph in your science notebook. Be prepared to present your group's ideas to the class.

2. After all the groups have shared their ideas, your teacher will use the ideas to devise a standard procedure. Record the procedure on Student Sheet 4.1: Finding the Density of Air.

3. Use the procedure to find the mass and volume of an air sample. Record your results and use them to calculate the density of air. Be sure to show your calculation.

4. Return the apparatus to the plastic box.

REFLECTING ON WHAT YOU'VE DONE

1. Your teacher will record each group's results for this experiment. Look at the results carefully. Answer the following questions on Student Sheet 4.1:

   A. How does the density of air compare with the density of solids and liquids?

   B. Are the results all the same? Do you think the procedure you used is very precise? Why do the class results vary so much?

   C. Based on your discovery that air does have density, why do you think some things float in air?

Balance

Rigid plastic bottle

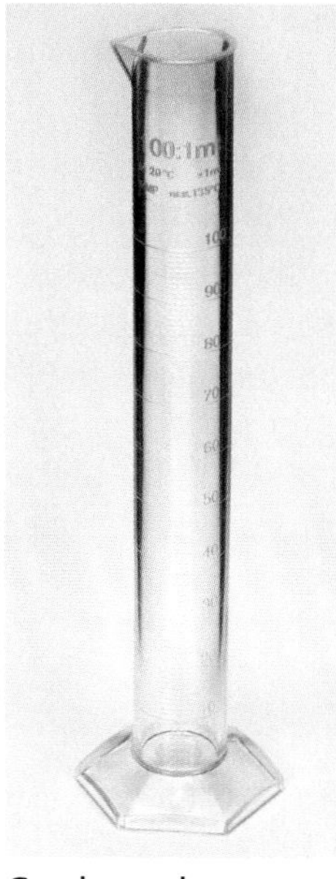

Graduated cylinder

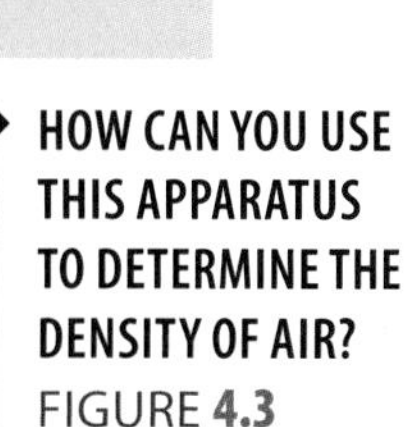

HOW CAN YOU USE THIS APPARATUS TO DETERMINE THE DENSITY OF AIR?
FIGURE **4.3**
PHOTOS: ©2009 Carolina Biological Supply Company

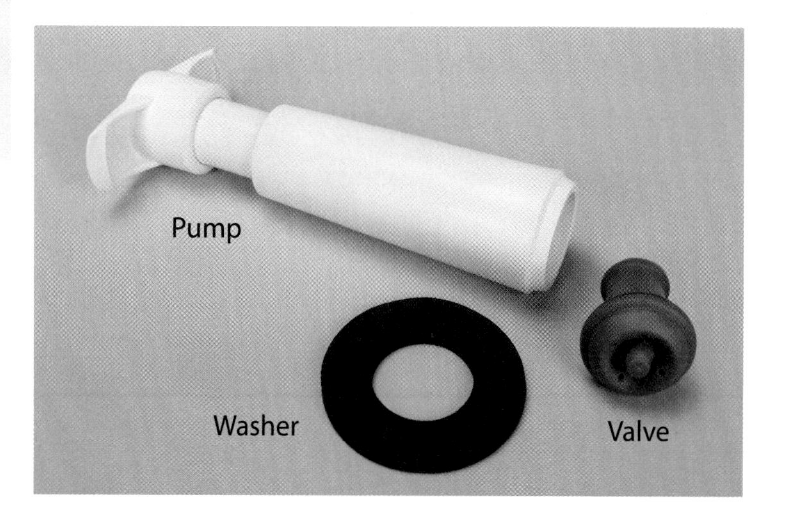

Pump

Washer

Valve

READING SELECTION

EXTENDING YOUR KNOWLEDGE

Deadly DENSITY

SOLDIERS WORE GAS MASKS DURING WORLD WAR I TO PROTECT THEMSELVES AGAINST POISONOUS GASES.

PHOTO: Library of Congress, Prints & Photographs Division, LC-USZ62-115014

Have you heard about a substance called chlorine? If you have, you probably know that it is sometimes added to water. Chlorine is added to drinking water to kill harmful microorganisms. When you go swimming, you can smell the chlorine in the pool. That's because chemicals that release chlorine are added to the water to keep it safe for swimming.

You may be surprised to learn that chlorine is a greenish-yellow gas. It is also a very poisonous substance. This property is exploited when chlorine is used as a disinfectant to kill microorganisms. In small amounts, chlorine kills microbes but not larger organisms. However, chlorine has also been used to kill people.

CHLORINE IS USED TO KILL THE MICROORGANISMS IN SWIMMING POOLS.

PHOTO: Jeff Sandquist/creativecommons.org

In World War I (1914–1918), chlorine was used as a weapon. Most of the battles in this war were fought between lines of trenches that provided the soldiers with some protection against gunfire. On April 22, 1915, at the battle of Ypres, in France, the Germans used a new secret weapon. That weapon was chlorine. They released chlorine gas from their side of the lines. The chlorine was carried by the wind to the enemy trenches. Because chlorine is much denser than air, it stayed near the ground and poured into the trenches. Choked and blinded, the defenders were then overrun by German troops wearing gas masks. After this gas attack, soldiers on both sides were issued gas masks. ■

DISCUSSION QUESTIONS

1. Carbon dioxide fire extinguishers are specialized to release carbon dioxide gas, which smothers fires by blocking oxygen from reaching them. What conclusion can you draw about the density of carbon dioxide relative to the density of oxygen?

2. Use library and Internet resources to find out more about the properties and uses of chlorine.

READING SELECTION

EXTENDING YOUR KNOWLEDGE

AIR HEADS

What do a scuba diver and an astronaut have in common? They both have air on their minds. Air is something most of us take for granted. We might think about it when we are swimming or exercising, but otherwise we know there is plenty of it around. The air that surrounds our entire planet is called the atmosphere. To an astronaut and a diver, air is something to think about. They have to carry an atmosphere with them: their own supply of air compressed into a small tank. If it runs out, they are in big trouble!

WHAT DO THESE TWO EXPLORERS HAVE ON THEIR MINDS?

PHOTO (left): Smithsonian Institution, Carl C. Hansen, Neg. #91-17588
PHOTO (right): NASA Headquarters—Greatest Images of NASA

Why do we need air? Sometimes people say, "We need air to breathe." This is the opposite of the truth. In fact, we breathe because we need air—even then, we only need part of it. Air is a mixture of gases. The part we use is called oxygen, and it makes up about one-fifth of normal atmospheric air.

Why is oxygen so important? Our bodies use oxygen to combine with food substances in a process called respiration. Respiration releases energy that we can use for our body processes. Respiration also makes carbon dioxide, which our bodies must release back into the air and water. If you were deprived of oxygen for more than a few minutes, your life processes would stop. You would suffocate to death!

A LAYER OF AIR CALLED THE ATMOSPHERE SURROUNDS EARTH.

PHOTO: NASA Headquarters—Greatest Images of NASA

READING SELECTION

EXTENDING YOUR KNOWLEDGE

THIS FIERCE FOREST FIRE WOULD BURN EVEN MORE EXPLOSIVELY IF THE ATMOSPHERE WERE PURE OXYGEN.

PHOTO: FEMA/Andrea Booher

While respiration takes place within our cells, burning processes outside of our bodies also require oxygen. Have you ever made a campfire? Think about how fanning the fire can help get it burning more quickly. Things burn even faster in pure oxygen. In fact, burning things in pure oxygen can be explosive. It's a good thing that air consists mainly of another gas called nitrogen.

From our "breathing" point of view, nitrogen gas doesn't do very much because our bodies don't use it in the process of respiration. Very few substances react with it, and it's colorless and odorless. Many of the other gases found in air don't do much either. Some of these gases, including argon, neon, and helium, are so renowned for doing nothing that they are called inert gases.

Other gases are more important to living things. Without the 0.03 percent of carbon dioxide in the air, there would be no green plants. Plants use carbon dioxide to make food. They use the energy from sunlight to combine water with carbon dioxide to make carbohydrates, which form building blocks for cells. Plants use parts of the air to create living matter! Most plants absorb water through their roots, but water is also found in the air (it's called water vapor). The amount of water in the form of vapor varies. On hot, sticky days, you can easily feel that there's a lot of water in the air.

THIS LEAF IS A FACTORY THAT USES CARBON DIOXIDE FROM THE AIR AS A RAW MATERIAL.

PHOTO: Criss!/creativecommons.org

READING SELECTION

EXTENDING YOUR KNOWLEDGE

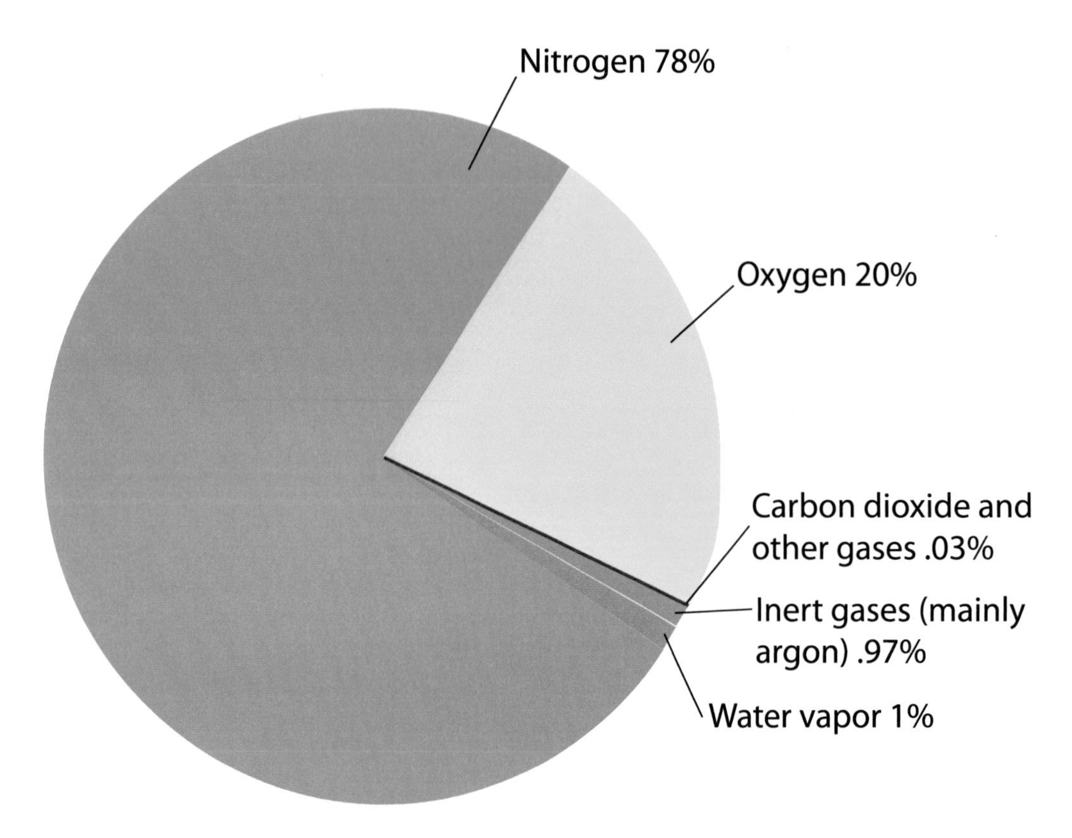

AIR IS COMPOSED OF SEVERAL DIFFERENT GASES. THIS CHART PROVIDES A SIMPLIFIED SUMMARY OF THE COMPOSITION OF AIR.

The amount of carbon dioxide and other gases in the air is affected by pollution. When oxygen is used in the burning of fossil fuels (oil, coal, and gas) for transportation, heating and cooling, and other energy supply needs, carbon dioxide is produced. The age of industrialization, which began about 150 years ago, has increased the amount of carbon dioxide in the air beyond its natural level. By the beginning of the new millennium (2000), it was significantly higher than at any time previous during the last 400,000 years!

Scientists have also noted increasing amounts of other gases in Earth's atmosphere. Methane, nitrous oxide, and other gases like carbon dioxide are by-products of the burning of fossil fuels. When they reach the atmosphere, these gases trap heat below (like the glass on a greenhouse), causing a warming effect by preventing it from escaping into space. This so-called "greenhouse effect" is responsible for rising temperatures (global warming) and other global changes including droughts, hurricanes, glacial melting, sea level rises, and animal habitat changes.

Scientists have been urging us to focus our scientific and technological efforts in the next century on reducing the production and impact of carbon dioxide and other greenhouse gases around the world. ■

DISCUSSION QUESTIONS

1. Use library or Internet resources to answer the following questions:
   A. Where did Earth's atmosphere come from, and how has its composition changed over time?
   B. What predictions do scientists make about how the composition of Earth's atmosphere will change in the future?
2. Do other planets in our solar system have atmospheres? If so, how do their atmospheres differ from Earth's?

LESSON 5

TEMPERATURE AND DENSITY

INTRODUCTION

Have you ever looked at a thermometer and wondered how it works? You may be surprised to learn that the thermometer was not invented until about 400 years ago. How did people measure temperature before that? Did they guess the temperature? Did they say something feels hot or cold? Thermometers are actually very easy to make from simple materials. In this lesson, you will build your own thermometer and learn something about how and why it works.

THIS THERMOMETER CONTAINS LIQUID. HOW DOES IT WORK?

PHOTO: National Science Resources Center

OBJECTIVES FOR THIS LESSON

- Use a thermometer and discuss the purpose of its different parts.
- Build a working thermometer and use it to measure temperature.
- Discuss how your thermometer works and relate this to changes in the volume and density of matter.

MATERIALS FOR LESSON 5

For you

1 copy of Student Sheet 5.1: Building a Thermometer
1 copy of Student Sheet 5.2: Replacing the Liquid with Air
1 copy of Student Sheet 5.3: Heating the Metal Strip
1 copy of Student Sheet 5: Homework for Lesson 5: Engineering for Expansion
1 pair of safety goggles

For you and your lab partner

1 thermometer
1 piece of plastic tubing mounted in a stopper
1 rubber stopper with single hole
1 20 × 150-mm test tube
1 250-mL beaker containing 100 mL of colored water
1 black permanent marker
1 metric ruler
Access to hot- and cold-water baths

GETTING STARTED

1. Listen carefully as your teacher reviews the safety tips. Then obtain the plastic box of apparatus for your group and check the items against the materials list. Each plastic box contains a set of apparatus for each pair in your group.

2. With your lab partner, closely examine the thermometer you have been given. Discuss with your partner the purpose of the different parts of the thermometer. The following questions will help you in your discussion:

   A. What do thermometers measure?

   B. Where is most of the liquid in your thermometer?

   C. What part of the thermometer do you place in the substance when you measure its temperature?

   D. What is the temperature range of your thermometer, and what units does it measure in?

   E. When you read the thermometer, which part of the liquid do you focus on?

   F. What do you notice about the distances between the marks on the scale?

3. Gently hold the bulb of the thermometer (the red end) in your hand. Discuss the following questions with your partner:

   A. What happens to the red liquid?

   B. What temperature does it reach?

   C. What happens to the reading when you let go of the bulb and hold on to the other end of the thermometer?

   D. What are you measuring when you release the bulb?

   E. Why do you think the liquid in the thermometer moves?

4. Be prepared to discuss your observations and ideas with the class.

SAFETY TIPS

Do not shake the thermometers. Unlike medical thermometers, the lab thermometers do not require "shaking down," and shaking increases the chance of breakage.

Always place the thermometers in a safe spot—do not let them roll off the table.

Handle the thermometers gently.

INQUIRY 5.1

BUILDING A THERMOMETER

PROCEDURE

1. Divide the contents of the plastic box between the two pairs in your group. How could you use the materials to build a thermometer? You have 5 minutes to discuss possible designs with your partner and draw your design on Student Sheet 5.1: Building a Thermometer. Do not build the thermometer yet.

2. Your teacher will ask different groups to share their designs with the class. The class will decide on the best design to use.

3. Follow the design discussed to build your thermometer.

4. Add a scale to the thermometer. This process is called calibration. To calibrate your thermometer, follow these instructions:

   A. Place the test tube end of the thermometer in the cold-water bath. Let it stand for about 5 minutes.

   B. Without removing the test tube from the cold-water bath, use the black permanent marker to make a mark on the plastic tubing at the water level.

   C. Use the thermometer in the water bath to record the temperature of the water bath.

   D. Place the test tube in the hot-water bath. Let it stand for about 5 minutes.

   E. Mark the tubing and record the temperature as before.

   F. Calculate and record the temperature difference between the two readings you made.

   G. What is the distance in millimeters (mm) between the two marks on the plastic tubing? Write your measurement under Step 3 on Student Sheet 5.1.

   H. Calculate the distance on your thermometer that is equal to 1°C.

   I. Use this information to figure out where 0°C and 100°C will be on your thermometer.

   J. Mark off the temperature scale between 0°C and 100°C in 5° intervals. Label these intervals every 10°C.

Inquiry 5.1 continued

5. Once you have built and calibrated your thermometer, test it by measuring room temperature. Allow time for your thermometer to reach room temperature. What reading did your thermometer give for room temperature? Record your answer.

6. Measure room temperature with the laboratory thermometer. What reading did the laboratory thermometer give? Record your answer.

7. Answer the questions in Steps 5-10 on Student Sheet 5.1.

INQUIRY 5.2

REPLACING THE LIQUID WITH AIR

PROCEDURE

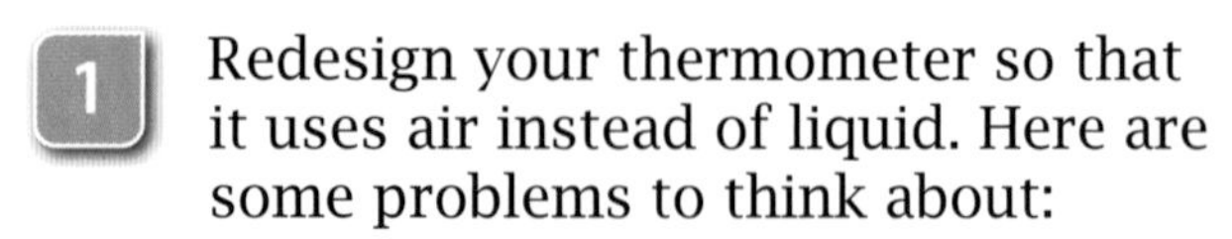

1. Redesign your thermometer so that it uses air instead of liquid. Here are some problems to think about:

   A. How will you stop the air from escaping?

   B. How will you measure the distance the air moves up the column?

2. Draw your design on Student Sheet 5.2: Replacing the Liquid with Air. After a short class discussion, build your thermometer and try to calibrate it.

3. Answer the following questions on Student Sheet 5.2:

   - What problems did you encounter when calibrating your air-filled thermometer?
   - How did the sensitivity of your air-filled thermometer compare with that of your liquid-filled one?

INQUIRY 5.3

HEATING THE METAL STRIP

PROCEDURE

1. Your teacher will pass around a metal strip. What do you notice about both sides of the strip?

2. What do you think will happen to the metal strip when it is heated? Write your prediction on Student Sheet 5.3: Heating the Metal Strip.

3. Observe what happens to the strip as your teacher heats it. Record your observations.

4. Observe the strip after it cools. Write a description of what happens.

5. Observe what happens to the strip when your teacher heats the other side of it. Record your observations. Why do you think the strip behaves this way? Write an explanation.

REFLECTING ON WHAT YOU'VE DONE

1. Answer the following questions in your science notebook and discuss your answers with your lab partner and the class:

   A. What do these three inquiries tell you about how the volume of matter is affected by temperature?

   B. How does the change in the volume of the air differ from the change in the volume of the liquid?

   C. How does this change in volume affect the density of solids, liquids, and gases?

   D. When measuring the density of a substance, why is it important to record the temperature of the substance?

   E. Are there any other uses for the expansion and contraction of matter?

   F. What problems could be caused by expansion or contraction of matter? (You may want to look at the reading selection about the Trans-Alaska Pipeline on pages 70–73 to help you answer this question.)

2. Read "Changing Temperature, Changing Density" on the next page.

READING SELECTION

BUILDING YOUR UNDERSTANDING

CHANGING TEMPERATURE, CHANGING DENSITY

Most matter increases in volume when it gets hotter. For example, if an iron rod is heated, it will get longer and fatter and its density will decrease. This happens because the mass of the rod stays the same, but its volume increases. The increase in the volume of matter with increasing temperature is called expansion. When cooled down, most matter decreases in volume and increases in density. This decrease in volume is called contraction.

A few substances behave differently when heated or cooled. Water is one such substance. When water approaches freezing, it expands and becomes less dense, which is why water pipes sometimes burst when they freeze and why icebergs float. ■

THIS HORSE THERMOMETER HAS A DIGITAL READOUT. VETERINARIANS CAN USE IT IN CONJUNCTION WITH A COMPUTER. IT DOES NOT USE CHANGING DENSITY TO MEASURE TEMPERATURE. DO YOU KNOW HOW IT WORKS?

PHOTO: Cynthia M. Green, GLA Agricultural Electronics

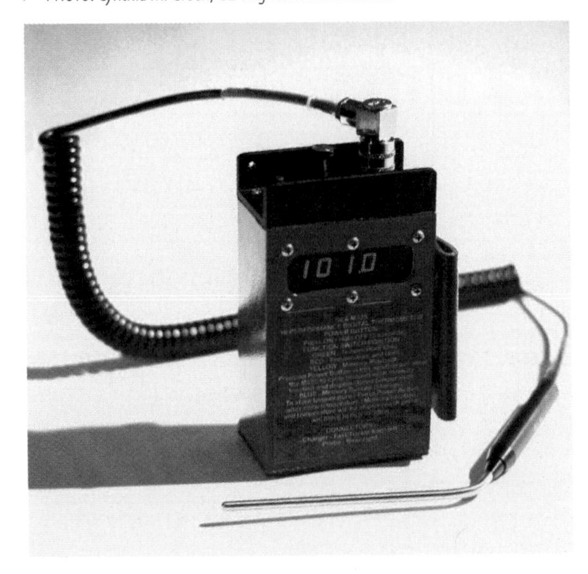

SOME THERMOMETERS USE BIMETAL STRIPS TO MEASURE TEMPERATURE. IN THIS LESSON YOU MAY GET SOME IDEAS ABOUT HOW THESE THERMOMETERS WORK.

PHOTO: Elektra Noelani Fisher/creativecommons.org

MEASURING TEMPERATURE by DEGREES

"How cold is it outside?" "Is your soup hot enough?" How many times have you been asked questions about temperature? Usually, we answer them according to how things feel to us. We compare temperatures with our own body temperature. People have always compared temperatures in this way. However, sometimes you need to know exactly how hot or cold something is. For example, if you cook a pizza in an oven that is too hot, it may burn—so you need to know the temperature of the oven.

About 400 years ago, some scientists began to tackle the problem of measuring temperature. Galileo was one of the first. He made a thermoscope. This was a device that could be used to compare temperatures. Look at the drawing of the thermoscope. Can you figure out how it worked?

It took another scientist, Olaus Roemer, a Dane who was interested in astronomy and meteorology, to come up with a way of comparing temperatures measured with different devices. In 1701, Roemer calibrated his temperature-measuring devices according to the temperatures of ice water and the human body. He had made the first thermometer.

A THERMOSCOPE BUILT TO ONE OF GALILEO'S DESIGNS. THERMOSCOPES WERE USED TO COMPARE TEMPERATURES BUT HAD NO STANDARDIZED SCALE.

READING SELECTION

EXTENDING YOUR KNOWLEDGE

GALILEO (1564-1642) BUILT SOME OF THE EARLIEST THERMOSCOPES, WHICH ARE THERMOMETERS WITHOUT SCALES.

PHOTO: Library of Congress, Prints & Photographs Division, LC-USZ62-7923

OLAUS ROEMER (1644-1710) INVENTED THE FIRST USEFUL TEMPERATURE SCALE.

PHOTO: Library of Congress, Prints & Photographs Division, LC-USZ62-124161

ANDERS CELSIUS (1701-1744) INVENTED THE CELSIUS, OR CENTIGRADE, SCALE.

PHOTO: Göran Henriksson, Department of Physics and Astronomy, Uppsala University

WILLIAM THOMSON (1824-1907) STARTED HIS TEMPERATURE SCALE WITH THE LOWEST POSSIBLE TEMPERATURE.

PHOTO: Library of Congress, Prints & Photographs Division, LC-USZ62-64292

Another scientist, this time from Holland, built on Roemer's ideas. His name was G. Daniel Fahrenheit. Fahrenheit altered Roemer's scale. He used the melting point of a salt-and-water slush as his zero point and the human body temperature as his high point. He divided the space between the two points into 96 degrees. The scale was later adjusted so that its calibration points were at 32°F for ice melting and 212°F for water boiling. On the adjusted scale, human body temperature became 98.6°F. The new scale was named after Fahrenheit and is still used today.

About 30 years later, in 1742, another scientist, Anders Celsius from Sweden, came up with a new scale. Celsius designated the melting point of ice as 100°C and the boiling point of water (at sea level) as 0°C. After Celsius's death, the scale was reversed so that the melting point of ice became 0°C and the boiling point of water (at sea level) became 100°C. This scale, called the Celsius (or centigrade) scale, was popular because it used two temperatures that most people easily understand. It's now used all around the world. This scale has one big problem. All temperatures below zero become negative numbers. Can you really have a negative temperature? Wouldn't it be better to start a scale at the lowest possible temperature and work your way up?

About 100 years later, in 1848, British physicist William Thomson could see the advantage of just such a scale. By that time, work done by Thomson and other scientists on how energy behaves in the universe led him to develop a scale that placed the absolute lowest possible temperature at zero. This temperature is the same as -273°C and is called absolute zero. An object at absolute zero contains its lowest possible energy. Scientists have come within a few billionths of a degree of reaching absolute zero in the laboratory. Thomson borrowed the divisions on Celsius's scale and made the melting point of ice 273 degrees. What happened to the Thomson scale? It is still used by scientists around the world, who consider it to be the most useful temperature scale.

Thomson was such a clever scientist and inventor that the British government made him a lord and gave him the title Lord Kelvin. So his scale became the Kelvin scale, and temperature is measured in kelvins (abbreviated as K). ■

DISCUSSION QUESTIONS

1. How does a degree on the Fahrenheit scale compare with a degree on the Celsius scale? Remember to use the calibration points on each scale in your comparison. Can you think of a way to change the temperature from one scale to the other?
2. How did later scientists improve upon the work of earlier scientists in the development of temperature scales?

READING SELECTION

EXTENDING YOUR KNOWLEDGE

Just a Load of HOT AIR

Just a load of hot air! That's what you will find inside a hot air balloon. But why does hot air help a balloon rise? Hot air balloons rely on the fact that the density of air decreases as it gets hotter. A gas burner, mounted above a balloon's basket, is used to heat the air inside the balloon. As the air heats, it increases in volume, or expands. You already know that density is equal to mass divided by volume. So what happens if the volume of the air inside a balloon increases while the mass of the air stays the same? The density of the air in the balloon decreases. When the average density of the balloon (including the burner, basket, and passengers) becomes less than the density of the surrounding air, the balloon begins to rise—it floats upward in the air.

The balloonist can alter the height of the balloon by switching the burner on and off. If the burner is turned off, the air inside the balloon cools. As the air cools, its volume decreases, it becomes denser, and the balloon goes down. The balloonist can also let some of the hot air out of the top of the balloon to make the balloon go down. If the burner is turned on, the air in the balloon becomes hotter. The air takes up more volume, becomes less dense, and the balloon rises.

The first free flight of a hot air balloon was in 1783, when the Montgolfier brothers sent a sheep, a duck, and a rooster into the air in a balloon made from linen. A few weeks later, in the first manned free flight, two Frenchmen, using burning straw as a heat source, piloted a Montgolfier balloon about 9 kilometers (about 5.6 miles) across Paris.

THE HOT AIR BALLOON RISES BECAUSE ITS AVERAGE DENSITY IS LESS THAN THE SURROUNDING AIR.

Today, most hot air balloons are made from ripstop nylon and use propane gas burners instead of straw. The average hot air balloon is as tall as a seven-story building, is about 20 meters (about 65.6 feet) across at the widest part, and is big enough to carry four adults. Hot air balloons can be built in many shapes. ■

HOT AIR BALLOONS CAN BE MADE INTO MANY FUN SHAPES, BUT WHY DO THEY NEED TO BE SO BIG?

PHOTO: Laura Rush/creativecommons.org

THIS IS A SCALE MODEL OF THE BALLOON USED FOR THE FIRST MANNED BALLOON FLIGHT. THE BALLOON LIKELY REACHED A HEIGHT OF MORE THAN 1,000 METERS (3,280 FEET) AND STAYED ALOFT FOR ABOUT 25 MINUTES.

PHOTO: San Diego Air and Space Museum

DISCUSSION QUESTIONS

1. How does a balloon pilot use density to control the altitude of a balloon?

2. A treat for tourists in Kenya is a hot air balloon ride over the Serengeti plains to observe many different species of animals. Why are these excursions generally scheduled for very early morning?

READING SELECTION

EXTENDING YOUR KNOWLEDGE

DENSITY CREATES CURRENTS

How do changes in density move matter? This movement involves a process called convection. To understand how convection works, imagine a room in a house, like the one shown in the picture below.

One side of the room has a heater; on the opposite wall is a window. On a winter day, when the heat is on, cold air near the heater will warm up. What happens to hot air? It expands, becomes less dense, and rises. On reaching the ceiling, it flows as drafts to the cooler parts of the room. The heated air starts to cool down the farther it drifts from the heater. This process is speeded up when the air meets the cold window. As the air cools, it becomes denser, sinks to the floor, and eventually completes a circuit of the room. A circular convection current is set up. Circular currents like this are called convection cells.

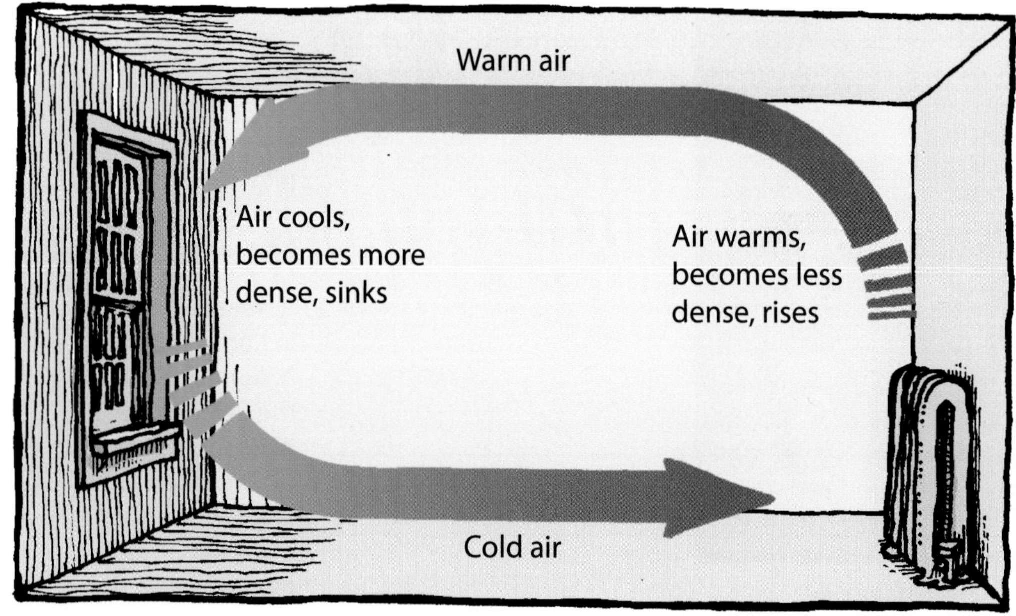

A CONVECTION CELL CAN OCCUR IN A ROOM.

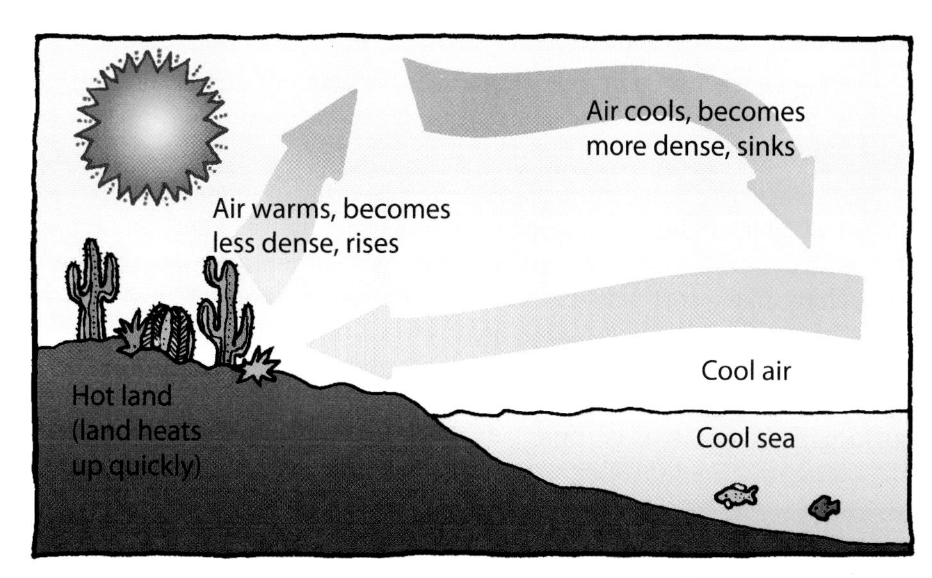

A CONVECTION CELL CAN ALSO OCCUR IN THE ATMOSPHERE.

Convection currents like this also take place in the atmosphere (see the picture on the top right). Air is heated near the equator and cooled at the poles. We encounter these convection currents as wind. Where do you think the heat energy for these convection currents comes from? Real winds are more complex than what is shown in the picture, but all winds are created by changes in density brought about by temperature differences. How do you think the winds shown in the picture would be different at night?

Convection works in liquids as well as in gases. Ocean currents have several different causes, many of which are due to changes in density. Some ocean currents are convection currents (see the picture on the bottom right).

Under the tropical sun, water at the equator warms up. At the cold poles, seawater cools down and sinks. Convection cells are set up with warm water moving along the surface to the poles and deep cold water flowing toward the equator. Changes in density caused by changes in salinity (the amount of salt in the water) are also important in the formation of ocean currents. For example, ice formation near the poles increases the salt content of the surface waters. The saltier the water, the denser it becomes. This colder, denser, more saline water sinks, creating its own density-driven currents. Surface winds can also set surface currents into motion.

THE GULF STREAM IS AN OCEAN CURRENT THAT IS DRIVEN, IN PART, BY CONVECTION. IT CARRIES WARM WATER, REPRESENTED IN ORANGE IN THIS IMAGE, FROM THE TROPICS TOWARD THE NORTH POLE, TRAVELING ALONG THE EAST COAST OF THE UNITED STATES.

PHOTO: NASA Goddard Space Flight Center Scientific Visualization Studio, Data provided by: Norman King (NASA/GSFC)

READING SELECTION

EXTENDING YOUR KNOWLEDGE

MOVING AND MAKING MOUNTAINS

Convection currents can move or split the oceanic or continental crusts. Radioactive elements deep within the earth provide the heat that drives these currents (see the picture at right illustrating how these convection cells work).

Earth's surface crust can be modeled as a series of giant rigid plates that fit together like a moving spherical jigsaw. These plates are called tectonic plates. They are composed of two types of crustal material: dense oceanic crust and less dense continental crust. The hot rocks deep in the mantle are solid but behave like a soft plastic. The "solid" mantle flows, ever so slowly, centimeters per year, like a very, very viscous liquid. When cooler, heavier rock on the surface of the earth sinks, the warm, less dense mantle rock is pushed to the surface. These convection currents create some of the mountain ridges found on the ocean bed. The Mid-Atlantic Ridge is one example. Sometimes these ridges emerge at the surface of the ocean as islands.

GREAT RANGES OF FOLD MOUNTAINS ARE FORMED WHERE PLATES COLLIDE. MOUNT EVEREST, THE HIGHEST MOUNTAIN IN THE WORLD, IS IN THE HIMALAYAS OF NEPAL. THESE MOUNTAINS ARE BEING FORMED AS THE PLATE CARRYING INDIA COLLIDES WITH THE PLATE CARRYING ASIA.

PHOTO: Rupert Taylor-Price/creativecommons.org

This continuous cycle of rising and sinking forms convection cells that cause the different plates to interact. As plates move, they push against other plates. When plates that consist of two pieces of continent push against one another, they may buckle up along their boundaries to form great fold mountains (see the photo of Mount Everest on the left). Tectonic plates may also slide past one another, as at the famous San Andreas fault in California.

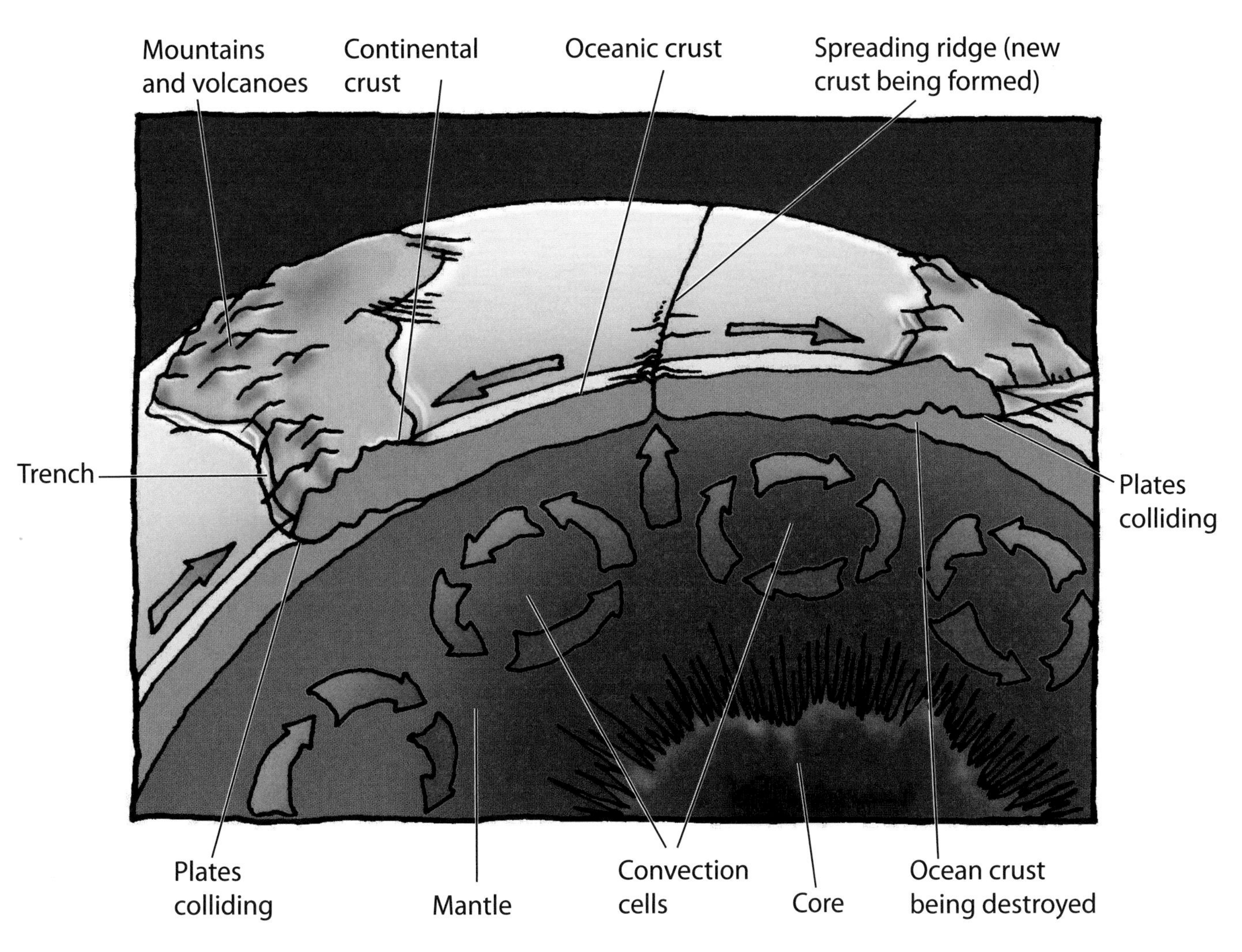

CONVECTION CELLS EXIST WITHIN THE EARTH.

READING SELECTION

EXTENDING YOUR KNOWLEDGE

CONVECTION CURRENTS IN THE EARTH PRODUCE VOLCANOES LIKE THIS ONE NEAR ICELAND IN THE NORTH ATLANTIC.

PHOTO: University of Colorado, National Geophysical Data Center, NOAA

Earthquakes can occur as the plates slide past one another or build mountains. If a plate of more dense oceanic crust pushes against less dense continental crust, what do you think happens? The more dense ocean crust sinks down to create ocean trenches. Evidence for this process is provided in the form of the volcanoes and earthquakes that are caused by all this activity.

Why is density important? Density differences and gravity are the same two forces that drive the convection cells in the atmosphere, oceans and solid earth. Next time you climb a mountain or hear about an earthquake or a tornado, think about how density and density changes have an impact on human lives! ■

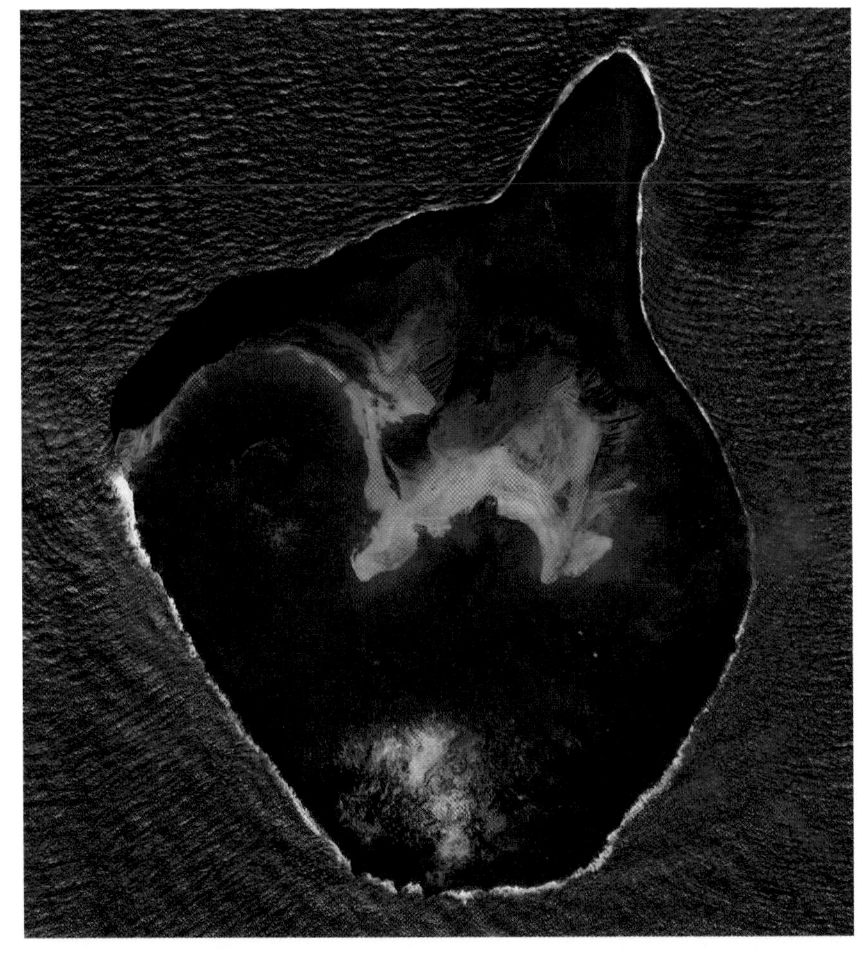

EVENTUALLY THIS VOLCANO FORMED AN ISLAND CALLED SURTSEY.

PHOTO: IKONOS satellite imagery courtesy of GeoEye. Copyright 2008.

SURTSEY IS PART OF THE MID-ATLANTIC RIDGE MOUNTAIN SYSTEM, FORMED AS A RESULT OF CONVECTION CURRENTS, THAT EXTENDS UNDER THE ATLANTIC OCEAN.

PHOTO: OAR/National Undersea Research Program (NURP)

DISCUSSION QUESTIONS

1. Look at the diagram of the room on page 64. If the heater were moved under the window, how might the movement of the air change?

2. Big birds such as vultures and hawks can often be seen gliding over big parking lots on sunny days, without even flapping their wings. How?

READING SELECTION

EXTENDING YOUR KNOWLEDGE

The Trans-Alaska Pipeline: MEETING NATURE'S CHALLENGES

The year was 1968, and the United States was concerned about its oil supply. With war brewing in the Middle East and an oil embargo threatening, where would the United States get the petroleum it needed? How could the country become less dependent on oil imports in the years ahead?

Just when concerns were getting serious, geologists discovered the largest oil field in this country—in Prudhoe Bay on the northern slope of Alaska. Part of the problem was solved.

But during the winter, the waters of Prudhoe Bay are frozen solid. For much of the year, they cannot be reached by sea-going oil tankers. How could those billions of gallons of oil be transported to the lower United States? The answer: Build a pipeline!

The people who took on this problem found themselves involved in one of the most difficult engineering challenges of the 20th century. To solve it, they had to focus on three features of the Alaskan territory: permafrost, earthquakes, and temperature extremes.

THE PIPELINE WAS MOUNTED ON POSTS ABOVE THE FROZEN GROUND. THE ALUMINUM RADIATORS ON TOP OF THE POSTS CONDUCT HEAT—LOST FROM THE PIPELINE—AWAY FROM THE SOIL. THE ZIGZAG IN THE PIPELINE ALLOWS IT TO EXPAND AND CONTRACT WITHOUT BREAKING.

PHOTO: U.S. Air Force photo by Staff Sgt. Joshua Strang

WATCHING OUT FOR PERMAFROST

At first, the engineers assumed that the pipeline would be buried underground. That's how most pipelines are built, after all.

But no one had ever built a pipeline in a place like Alaska, where it gets so cold that in many parts of the state, the subsoil is permanently frozen. This deep soil, which never thaws, is called permafrost.

Planners realized that the pipeline couldn't be buried in the permafrost because the heat of the oil, which comes out of the ground at about 77°C (about 170.6°F), could cause the icy soil to melt. If the icy soil melted, the pipe would sag and it might leak. In winter, the soil around the pipe would freeze again. This freeze-thaw cycle could cause the pipe to move enough to cause serious damage.

To avoid these complications, the engineers made an important decision: About one-half of the pipeline (about 700 kilometers or 435 miles) would have to be built above ground. They supported the pipe with posts that are topped with aluminum radiators. The posts conduct heat away from the soil, and the radiators transfer the heat to the surrounding air. The pipeline is also wrapped in 10 centimeters (3.9 inches) of fiberglass insulation. Both of these measures help to keep the permafrost solid.

FIBERGLASS INSULATION IS BEING WRAPPED AROUND THE PIPELINE TO REDUCE HEAT LOSS.

PHOTO: Courtesy of Alyeska Pipeline Service Company

READING SELECTION

EXTENDING YOUR KNOWLEDGE

FACTS AND FIGURES

- The pipeline is 1,287 kilometers (800 miles) long. Each piece is 122 centimeters (48 inches) in diameter.
- The pipeline crosses 34 major rivers and streams and 3 mountain ranges.
- Construction was started in 1973 and completed in 1977. The cost was $8 billion.

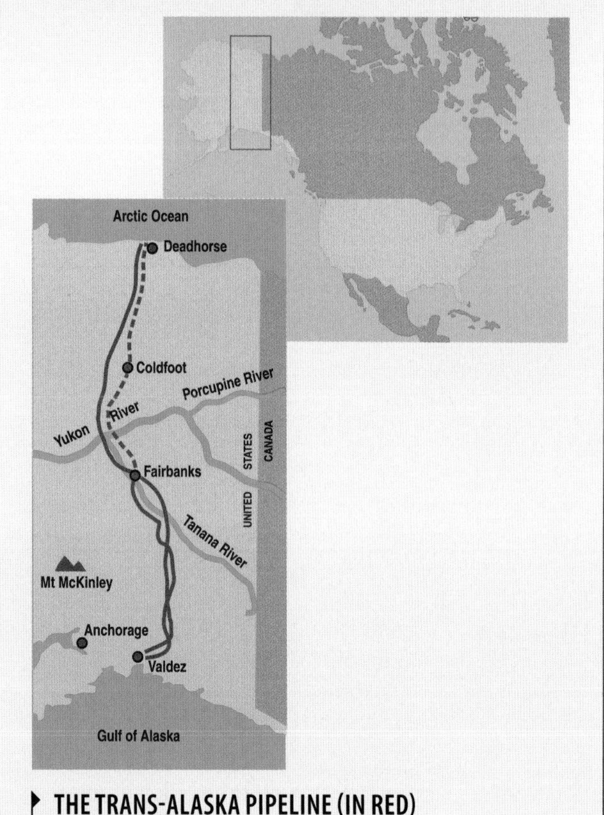

THE TRANS-ALASKA PIPELINE (IN RED) STRETCHES FROM VALDEZ TO DEADHORSE.

BLOWING HOT AND COLD

A second challenge was Alaska's temperature, which ranges between -60˚C (-76˚F) and 35˚C (95˚F). Because the metals from which the pipeline is made expand and contract with changes in temperature, the pipeline had to be built to accommodate changes in length. The engineers estimated that a 304-meter (997-foot) segment of pipeline could shrink by as much as 0.3 meter (1 foot) in the coldest weather and expand by an equal amount during the warmest season. That doesn't sound like much of a change, unless you realize that the pipeline is nearly 1,500 kilometers (932 miles) long! If the pipeline were straight, even a small change in each segment of the pipeline would be disastrous. The pipeline would either snap if it contracted too much or buckle if it expanded.

To prevent the pipeline from breaking, the designers used a zigzag configuration. These bends help relieve the effect of contraction and expansion.

ACCOUNTING FOR EARTHQUAKES

As if these extreme temperatures weren't enough, engineers had to deal with another big problem: earthquakes. Earthquakes are fairly common in Alaska. In fact, the largest earthquake ever to occur in the United States (measuring 9.2 on the Richter scale) took place in southern Alaska. The engineers had to build a pipeline that could survive such an event intact.

They designed a two-part system of "shoes" and "anchors" that hold the pipeline in place at weak areas (faults) where earthquakes have occurred, yet allow it to move enough so that it does not fall off its supports if the ground moves. At the Denali fault zone, where earthquake activity has been heavy, the pipeline is designed to move up to 6 meters (19.7 feet) side to side and 1.5 meters (4.9 feet) up and down. ■

DISCUSSION QUESTIONS

1. How did engineers overcome the challenge of a 95-degree temperature range when designing the Trans-Alaska Pipeline?
2. The pipeline connections to its above-ground supports allows for a small amount of side-to-side motion. Explain how these connections and the zigzag design of the pipeline relieve the strains caused by thermal expansion and contraction.

LESSON 6

APPLYING THE HEAT

INTRODUCTION

Have you ever baked a chocolate cake? After mixing the ingredients, you have a sticky brown liquid. You put it in the oven at a certain temperature. Forty to 50 minutes later . . . presto! You have a chocolate sponge cake. Now for the chocolate topping. You slowly heat the chocolate until it melts and add it to the rest of the frosting ingredients. Then you spread it quickly on top of the cake before it turns solid again. The next step? Well, eat it, of course!

EVEN BAKERS USE HEAT TO CHANGE MATTER.

PHOTO: U.S. Navy photo by Mass Communication Specialist 3rd Class Paul J. Perkins.

It's time to stop thinking about food and get on with the science. Of course, what you just read contains a great deal of science. Most of the ingredients in a cake change when they are heated. But each one changes in a different way. The way they behave depends on their characteristic properties.

The changes that occur when a cake mixture is heated are very complex. Some of the substances in the mixture change phase; others break down or combine with one another to form new substances. It is easier to examine the effect of heating on other, simpler substances. In this lesson, you will discuss how heating affects some familiar substances. Next, you will investigate six different substances as they are heated. You will be asked to make careful observations, accurately record them, and discuss your results.

OBJECTIVES FOR THIS LESSON

- Review what you already know about how heating affects substances.
- Observe and record the effects of heating on different substances.
- Discuss the results of the inquiry.

MATERIALS FOR LESSON 6

For you

1 copy of Student Sheet 6.1: Applying the Heat
1 pair of safety goggles

For your group

1 burner
1 250-mL beaker
1 test tube clamp
5 test tubes
1 lab scoop
1 test tube brush
1 test tube containing sulfur
5 jars containing:
Ammonium chloride
Copper (II) carbonate
Copper (II) sulfate
Sodium chloride
Zinc oxide

GETTING STARTED

1. Take 5 minutes to think of two familiar household substances that you have heated. In your science notebook, write what happened when you heated the two substances.

2. Your teacher will lead a brainstorming session on heating substances. Be ready to contribute your examples and ideas to the discussion.

SAFETY TIPS

Use safety goggles at all times.

Tie back long hair and restrict loose clothing.

Never smell or taste chemicals.

Handle chemicals only with the lab scoop.

Use only a test tube clamp to pick up test tubes.

Never move around with a lit burner.

Never refill the alcohol burners.

Do not walk around while substances are being heated; remain at your workstation at all times.

Follow classroom procedures for disposing of broken glassware and cleaning up spills.

Wash your hands at the end of the lesson.

INQUIRY 6.1

HEATING SUBSTANCES

PROCEDURE

1. Your teacher will explain the purpose of Inquiry 6.1.

2. Your teacher will go over the safety procedures you will follow for heating substances. Listen carefully.

3. Watch the demonstration of the proper procedure for this inquiry. Your teacher will ask for one or two student volunteers to help with the demonstration. Ask questions about anything that isn't clear to you.

4. Participate in a class discussion on the procedure you will follow in this inquiry.

5. Copy the information from the class table onto Table 1 on Student Sheet 6.1: Applying the Heat. When you are finished, get into your lab group and collect the plastic box containing the apparatus. Check the contents of the plastic box against the materials list.

6. Read Steps 7 through 17 before starting to heat the substances.

7. Your teacher will designate a workstation for your group. Each workstation has a burner. Check that it is ready to use before asking your teacher to ignite it. Make sure you carefully follow the procedure demonstrated by your teacher for using your burner.

8. Place one lab scoop of the first substance into a test tube.

9. Write the name of the first substance you are going to heat in the first column of Table 1 on Student Sheet 6.1.

10. Examine the substance carefully. In the second column of Table 1, record its appearance before you start heating it.

11. Attach the test tube clamp near the mouth of the test tube.

Inquiry 6.1 continued

12. Heat the bottom of the test tube containing the substance for 1-2 minutes. Keep the tube moving to heat the substance evenly (see Figure 6.1).

SAFETY TIPS

Do not walk around while holding the hot test tube.

Hold the test tube at an angle of about 45 degrees to the flame. Make sure you are not pointing the open end of the test tube at yourself or anyone else.

13. Observe any changes and record your observations in the third column of Table 1.

14. Allow the test tube and the substance to cool for 1 minute. Place the test tube in the 250-mL beaker. In the fourth column of Table 1, record the appearance of the solid after it has cooled.

15. Use the same procedure to heat the other four substances in the jars.

16. Use the same procedure to observe, describe, and heat the sulfur. After cooling, return the test tube containing the sulfur to the plastic box. Another class may use these test tubes later.

17. When you finish heating all six substances, extinguish the burner.

18. Make sure all the tubes are cool before disposing of the chemicals in the container provided for that purpose. Do not empty or clean the test tube containing the sulfur.

19. Use the test tube brush and water to clean the other test tubes thoroughly. Allow the tubes to drain and then stand them upside down in the 250-mL beaker. Return your apparatus to the plastic box.

20. Answer the questions in Steps 2-4 on Student Sheet 6.1 and be prepared to discuss your answers with the class.

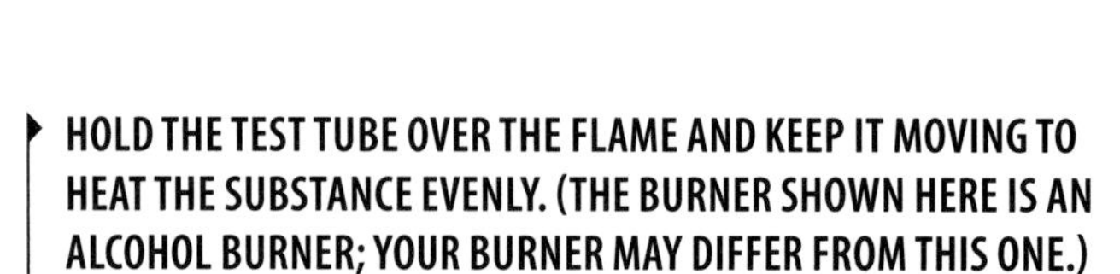

HOLD THE TEST TUBE OVER THE FLAME AND KEEP IT MOVING TO HEAT THE SUBSTANCE EVENLY. (THE BURNER SHOWN HERE IS AN ALCOHOL BURNER; YOUR BURNER MAY DIFFER FROM THIS ONE.)
FIGURE **6.1**

READING SELECTION

BUILDING YOUR UNDERSTANDING

HEAT AND CHANGING MATTER

You have discovered that heat affects different substances in different ways. The way a substance behaves when it is heated is a characteristic property of that substance. When you heated one of the solids, it melted. Then when the substance cooled, it solidified again. Another substance turned into a gas and then turned back to a solid farther up the tube. When a substance changes from one phase to another it is called a phase change. Did any of the substances you heated exist in all three phases of matter (solid, liquid, and gas) within the test tube?

Many of the other substances changed their appearance when you heated them and did not return to their original form when they cooled. This is usually a sign that a chemical reaction has taken place. In a chemical reaction, one or more substances (called reactants) are changed into new substances (called products).

In Inquiry 6.1, the substances that underwent chemical reactions decomposed when they were heated and then formed products that did not look like the original substances. This type of chemical reaction is called thermal decomposition. Some of the substances you heated decomposed and gave off invisible gas as one of the products. Can you identify one of the substances that you heated that did this? What evidence do you have that an invisible gas may have been produced? There are many other types of chemical reactions. You will investigate some later in the unit.

Some substances change in other ways when they are heated. Can you identify any of these changes? How could these types of changes be useful to people? ■

REFLECTING ON WHAT YOU'VE DONE

1. Participate in a class discussion about the changes you observed.
2. Read "Heat and Changing Matter."

READING SELECTION

EXTENDING YOUR KNOWLEDGE

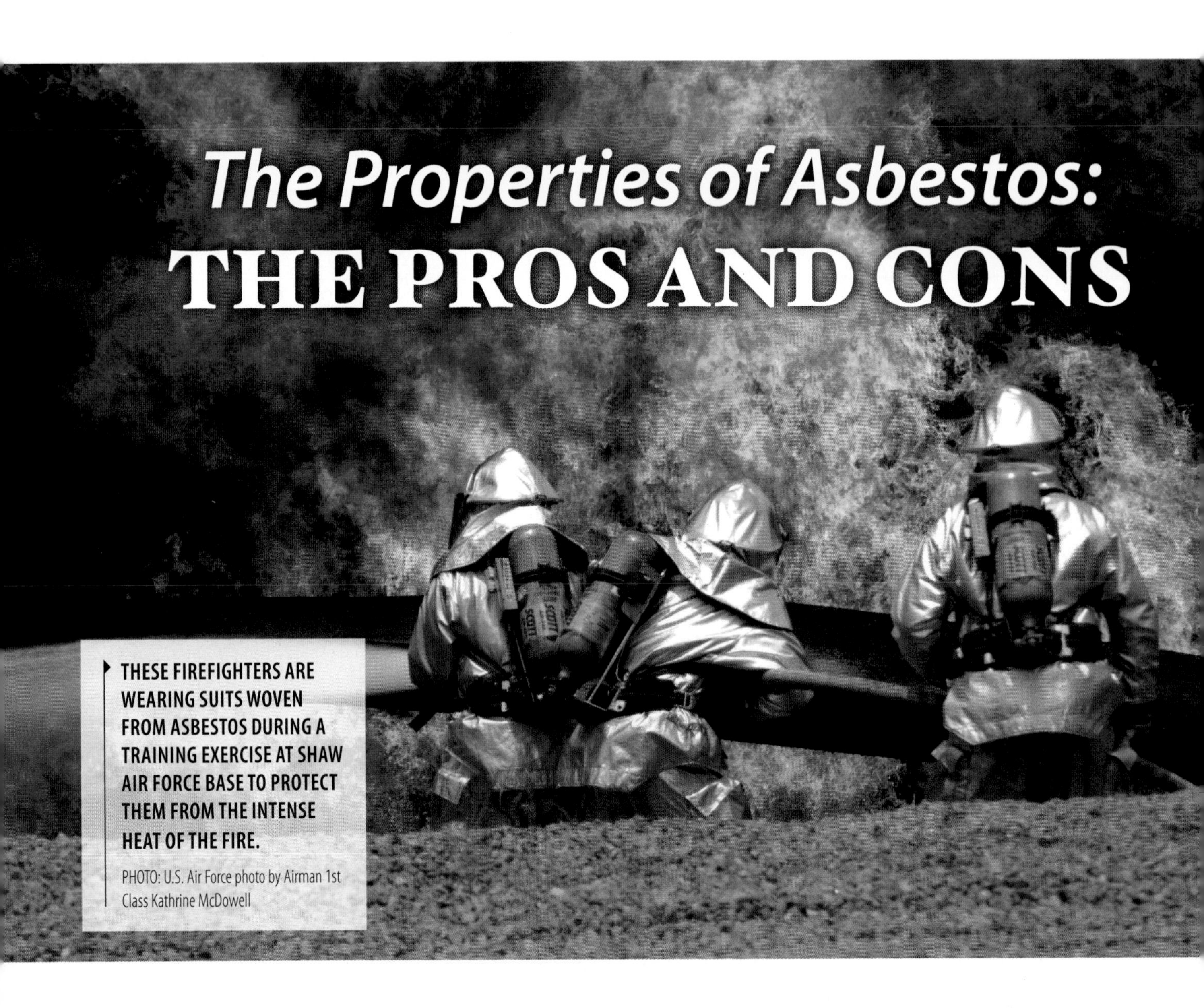

The Properties of Asbestos: THE PROS AND CONS

THESE FIREFIGHTERS ARE WEARING SUITS WOVEN FROM ASBESTOS DURING A TRAINING EXERCISE AT SHAW AIR FORCE BASE TO PROTECT THEM FROM THE INTENSE HEAT OF THE FIRE.

PHOTO: U.S. Air Force photo by Airman 1st Class Kathrine McDowell

Many substances burn when they are heated. Others melt or evaporate. Some substances, such as asbestos, do not change when they are heated. This property can be very useful. For centuries, people have known that this fibrous mineral has many useful properties. It is fire resistant. It does not melt or react with air, at least not until it gets very hot. One form of the mineral withstands temperatures up to 2750°C (4982°F). It is a very good insulator. It is strong. It resists acid. It is chemically inactive. It can be woven into cloth. Asbestos has some very useful properties, and it is readily available at a low cost.

The Romans used asbestos for lamp wicks. Egyptians used it to make burial cloths. In modern times, asbestos has been used in roofing and flooring, electrical and heat insulation, and brake linings. Because of its fire-resistant properties, asbestos has been used for a wide variety of other purposes, from theater curtains to firefighters' suits and gloves.

Until the 1970s, asbestos was widely used and asbestos mining and production were important industrial activities in the United States. Today, asbestos mining is banned in this country, and the use of asbestos has been strictly regulated.

Why? It is now known that inhaling asbestos fibers can cause lung disease. Asbestos releases tiny particles that remain suspended in the air. Once inhaled, these downy particles can remain in the lungs for decades. They cause delicate lung tissue to stiffen. A lung disease, called asbestosis, and a type of cancer may occur years after the original exposure.

ASBESTOS FIBERS AS SEEN THROUGH A MICROSCOPE.

PHOTO: U.S. Geological Survey

THIS IS A CLOSE-UP OF A PIECE OF ASBESTOS ROCK. CAN YOU SEE THE FIBERS?

PHOTO: Andrew Silver, Mineral Collection of Brigham Young University Department of Geology, Provo, Utah/U.S. Geological Survey

READING SELECTION

EXTENDING YOUR KNOWLEDGE

Today, construction companies are not allowed to use asbestos as insulation or fireproofing in new buildings. Workers who are exposed to asbestos must wear protective clothing. They have to shower and change clothes before going home.

Government regulations also apply to some buildings that had already been built when the new laws were passed. For example, schools that contain asbestos products have had to remove them.

ASBESTOS IS A NATURALLY OCCURRING MINERAL THAT IS MINED, PRIMARILY BY STRIP-MINING THAT LEAVES THE GROUND BARE.

PHOTO: LHOON/creativecommons.org

So, there's good news and bad news. Regulating the mining, manufacture, and use of asbestos has reduced the health risk that millions of Americans were being exposed to daily. But nothing has yet been found that can replace asbestos. However, researchers are exploring the use of synthetic fibers, fiberglass, and plastics as asbestos substitutes.

It's a trade-off: using a substance that has many useful properties versus having a safer environment. In the United States, the decision has been made. What other similar trade-offs can you think of? ■

ASBESTOS IS STILL PRESENT IN SOME BUILDINGS. HOW DO WORKERS IN THOSE AREAS MINIMIZE THE RISKS OF EXPOSURE?

PHOTO: Adrien Lamarre/U.S. Army Corps of Engineers

ASBESTOS FIBERS LODGED IN THE HUMAN LUNG (SUCH AS THE ONE SHOWN HERE) CAN CAUSE SEVERAL DISEASES, INCLUDING BLACK LUNG. OFTEN, THESE DISEASES ARE FATAL.

PHOTO: Centers for Disease Control and Prevention/Dr. Edwin P. Ewing, Jr.

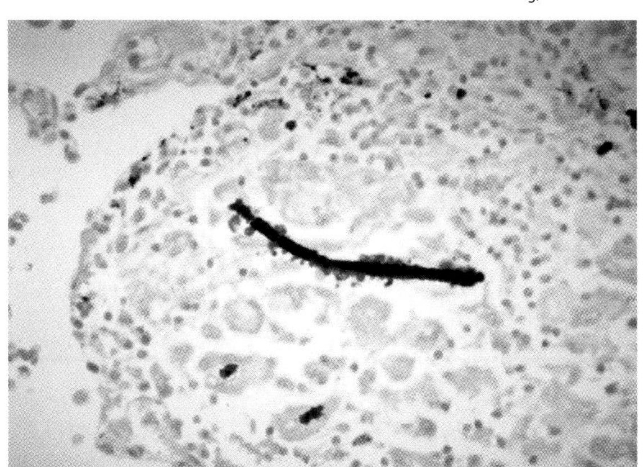

DISCUSSION QUESTIONS

1. What has asbestos been used for historically? What other materials might meet those needs?

2. Many substances with useful properties have undesirable ones as well. Use library and Internet resources to research some of the pros and cons of using mercury, plutonium, and benzene.

LESSON 7

JUST A PHASE

HOW ARE HEAT AND PHASE CHANGE INVOLVED IN POWERING THIS OLD LOCOMOTIVE?

PHOTO: Library of Congress, Prints & Photographs Division, LC-USZ62-72119

INTRODUCTION

Do you eat spaghetti? To cook spaghetti, you fill a large pot with water, perhaps add some salt, put the pot on the stove, and turn on the heat. When the water boils, you add the spaghetti, stir it well, and simmer it for about 8 minutes. Why do you need to cook it for 8 minutes? What happens if you are in a hurry and want to cook it faster? Will turning up the heat raise the temperature and make the spaghetti cook faster? Does the boiling water get hotter the longer it boils? Why don't the cooking instructions tell you to cook the spaghetti as quickly as you can?

You may be surprised to learn that a lot of interesting science takes place when you heat water. In this lesson, you will look at and think about what happens when solid water, or ice, is heated until it boils, and beyond.

OBJECTIVES FOR THIS LESSON

- Discuss your current knowledge of phase change.
- Observe what happens to ice as it is heated.
- Measure the temperature of ice water as it is heated.
- Plot a graph of your measurements.
- Interpret your graph and other observations.
- Relate the change in phase to the kinetic molecular theory of matter.

MATERIALS FOR LESSON 7

For you

1	copy of Student Sheet 7.1: Heating Ice Water
1	pair of safety goggles

For your group

1	burner
1	burner stand
1	thermometer
1	250-mL beaker
1	plastic soda bottle
15–20	craft beads
3–4	ice cubes (or crushed ice)
	Access to a clock or watch with a second hand
	Access to water

GETTING STARTED

1. After reading the Introduction, discuss the following questions with your group:

   A. How could you make spaghetti cook faster?

   B. Why does ice melt?

   C. Why can you play in the snow when it is warm outside?

   D. Why doesn't ice melt immediately when you add it to a soft drink?

   E. Are things that are boiling always hot? Are things that are frozen always cold?

   F. What are some of your own questions about what happens when ice melts and water boils?

2. Record your group's and your own ideas in your science notebook. You will be asked to present some of these ideas during a short brainstorming session.

CAN YOU MAKE SIMMERING SPAGHETTI COOK FASTER BY TURNING UP THE HEAT?
FIGURE **7.1**

INQUIRY 7.1

HEATING ICE WATER

PROCEDURE

1. Your teacher will explain how you should heat some ice. Watch and listen carefully.

2. Your teacher will designate a heating station for your group. Decide which job each member of your group will do: One of you will measure the temperature, one will monitor the time, and two will record the temperature and make observations.

SET UP YOUR APPARATUS AS SHOWN. (THE BURNER SHOWN IN THIS PICTURE IS AN ALCOHOL BURNER. YOUR BURNER MAY DIFFER FROM THIS ONE.)

FIGURE **7.2**

PHOTO: ©2009 Carolina Biological Supply Company

3. Carefully read the following instructions (A through L) and the safety tips on page 88.

   A. Go to your heating station.

   B. Check the apparatus against the materials list.

   C. Pour cold tap water to a depth of 1 cm into the bottom of the beaker.

   D. Obtain three or four ice cubes for your beaker. If the ice is crushed, get enough to fill the beaker to a depth of about 3 cm.

   E. Place the thermometer in the ice water and allow it to stand for a few minutes.

   F. Ask your teacher to light your burner. (If it is a Bunsen burner, adjust the flame so it is about 5 cm high.) Do not start heating the ice water yet.

   G. Make sure your apparatus is set up as shown in Figure 7.2.

Inquiry 7.1 continued

HEAT THE ICE AS SHOWN. MEASURE AND RECORD THE TEMPERATURE EVERY 30 SECONDS.
FIGURE **7.3**
PHOTO: ©2009 Carolina Biological Supply Company

H. Measure the temperature of the ice water (by this time, some of your ice will have melted). Record the temperature next to time 0 in Table 1 on Student Sheet 7.1: Heating Ice Water.

I. Put the beaker on the burner stand (see Figure 7.3). Slide the stand gently over the burner. Do not adjust the flame of your burner while you are heating the ice. Immediately start measuring the time.

J. Take the temperature every 30 seconds and record it. Observe any changes that take place and record these observations in Table 1.

K. When the water boils, continue to take readings for 3 minutes more.

L. Extinguish the burner and allow the apparatus to cool.

SAFETY TIPS

Tie back long hair.

Wear safety goggles throughout the inquiry.

Be careful when you handle hot apparatus.

4 Follow Steps A through L to perform the inquiry.

5 When you finish, return to your desk and make sure all members of your group have a complete copy of all the results and observations. Use this opportunity to discuss the observations you made with your group.

6 If your apparatus has cooled, pour out the water and clean up as instructed by your teacher.

7 Plot your data on the graph paper on Student Sheet 7.1. Give your graph a title. Decide which of the axes you will use for time and which for temperature. Remember, the independent variable goes on the horizontal (x) axis. If you are unsure what to do, discuss with your teacher how to lay out your graph. Make sure you accurately plot all of the points before you connect them to draw a curve. Label the curve to show where important changes took place (for example, "All the ice melted").

REFLECTING ON WHAT YOU'VE DONE

1. Discuss the following questions with your group:

   A. How does the shape of your curve compare with those produced by other groups?

   B. Do any changes in the direction of your curve match the point at which the ice melted or the water boiled?

   C. How can you use the curve on your graph to determine the temperature at which ice melted and water boiled? Are these temperatures what you expect?

2. Record your melting and boiling points on your graph.

3. Share your graph with the class, and your ideas about melting and boiling points.

4. On Student Sheet 7.1, make a copy of the diagram drawn by your teacher and label it. Write a paragraph that explains the shape of the curve on your graph.

5. Listen as your teacher explains the Kinetic Molecular Theory, or the Moving Molecule Theory of matter, which scientists now use to explain the phase changes you have observed.

6. Observe your teacher's demonstration of kinetic energy in different phases of matter. Use your own bottle and beads to demonstrate energy in different phases of matter.

7. In your science notebook, explain the kinetic theory of matter and illustrate the different phases of matter.

READING SELECTION

EXTENDING YOUR KNOWLEDGE

BOILING OIL

The next time you visit a gas station, look at some of the products sold there. In addition to gasoline, you'll probably find diesel fuel, kerosene, engine oil, gear grease, and other lubricants. All of these substances help keep cars and other vehicles running. Where do they come from?

A car's ability to cruise down the highway—and possibly even the highway itself—is based on the activity of a variety of organisms that lived in the sea hundreds of millions of years ago. These organisms died, but they only partially decomposed. Over millions of years, their remains became compressed, were heated, and eventually turned into crude oil and natural gas. By drilling down into the rocks where the crude oil and gas are found, it is possible to extract it.

THIS OLD TEXAS GUSHER SHOWS CRUDE OIL, UNDER PRESSURE FROM SURROUNDING GAS, BEING FORCED UP FROM UNDER THE GROUND. ONCE OUT OF THE GROUND, THE OIL MUST BE PROCESSED.

PHOTO: Library of Congress, Prints & Photographs Division, LC-USZ62-54453

OIL IS A VERY VALUABLE RESOURCE, AND PEOPLE GO TO GREAT LENGTHS TO EXTRACT IT. THIS DRILLING PLATFORM IN THE OCEAN IS DESIGNED TO EXTRACT VALUABLE OIL FROM THE EARTH.

PHOTO: National Oceanic and Atmospheric Association

Crude oil is very thick, comes in a variety of unusual colors (including red), and smells pretty awful. How is crude oil made into substances that can be used in cars? When crude oil is boiled, a lot of interesting things happen. You see, crude oil is not a single substance. It's a mixture.

As crude oil is heated, some of the substances in it start to boil. Substances with low boiling points, like gasoline, are the first to boil. Kerosene is next, followed by fuel oils (some of which are used to make diesel fuel) and then heavier oils (which are often used for lubrication). In the end, only the substances with high boiling points (more than 370°C, or 698°F) are left—as a dark, gooey, smelly mess. But even these materials have a use. They make up asphalt, which is used as a surface for the highways on which cars travel.

ALL OF THESE DIFFERENT OILS (OR FRACTIONS) WERE OBTAINED FROM CRUDE OIL BY THE PROCESS OF FRACTIONAL DISTILLATION. EACH FRACTION HAS A DIFFERENT BOILING POINT.

PHOTO: Courtesy of Pennsylvania Historical & Museum Commission, Drake Well Museum, Titusville, PA

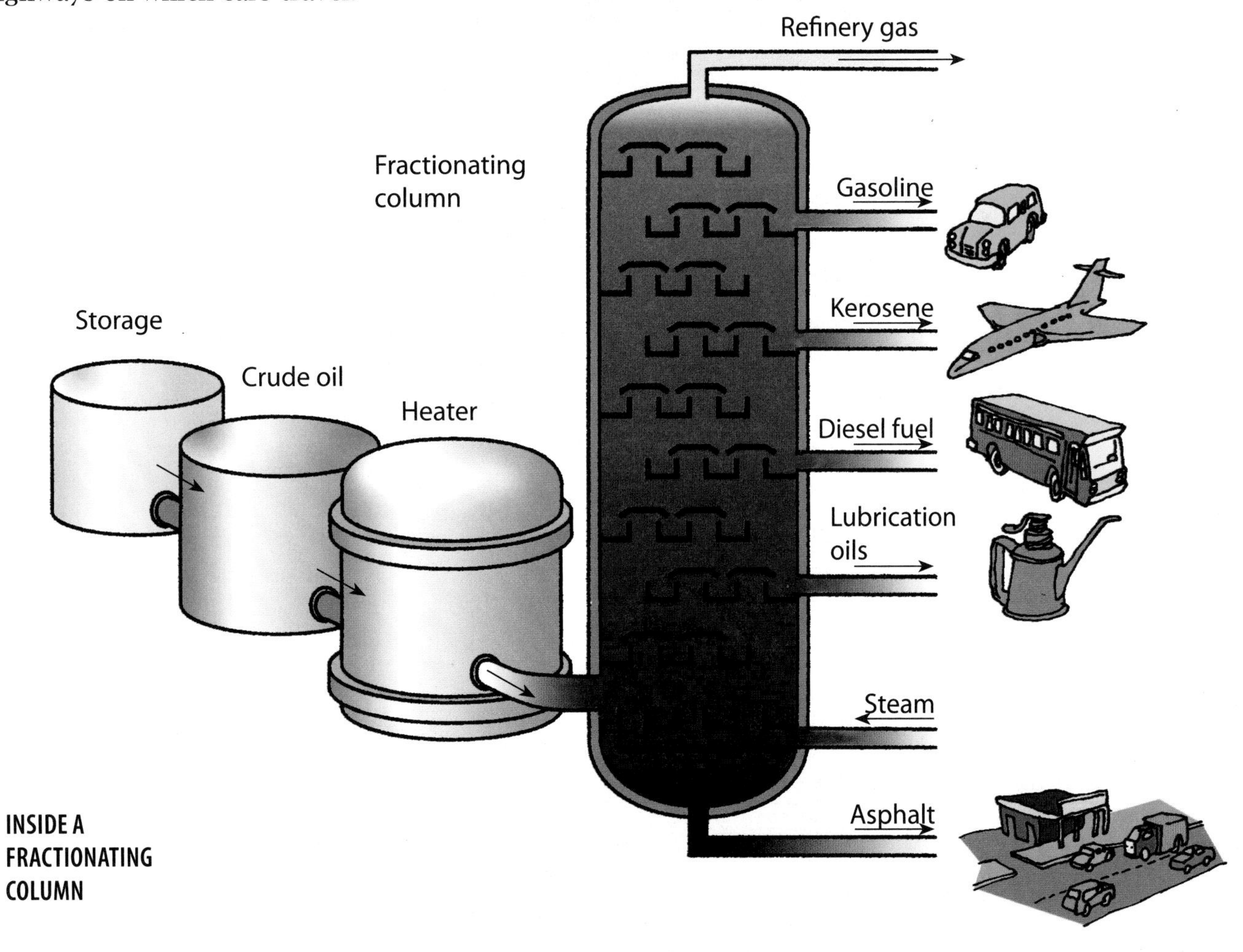

INSIDE A FRACTIONATING COLUMN

READING SELECTION

EXTENDING YOUR KNOWLEDGE

All this boiling takes place at an oil refinery. The crude oil is heated in special towers called fractionating towers or columns, which contain steel trays for separating and collecting the various liquids that result from boiling. In different levels of the towers, different gases are produced when different substances within the crude oil (like gasoline or kerosene) begin to boil. These gases then condense onto the trays. The substances that come from crude oil are not only used to make fuels and lubricants. They are also used as raw materials for hundreds of industrial processes—from making glues to plastic bottles and even clothing. ■

CRUDE OIL IS SEPARATED INTO ITS COMPONENT PARTS IN FRACTIONATING TOWERS LIKE THESE.

PHOTO: Courtesy DOE/NREL, Credit — David Parsons

DISCUSSION QUESTIONS

1. Examine the illustration at left. How many items in it, including car parts, could be made from oil?
2. Crude oil is converted into various substances that we use for different purposes. What is another natural resource that is converted into various useful substances?

LOST WAX CASTING:

EXPLOITING MELTING POINTS FOR ART AND INDUSTRY

THE WAX MODEL IS CAREFULLY CLEANED BEFORE THE FIRST LAYER OF FINE CLAY IS ADDED.

PHOTO: Eliot Elisofon, Photographic Archives, National Museum of African Art, Smithsonian Institution, EEPA 6963

Knowledge of melting points is very important for people who work with metal. Let's look at the goldsmiths who live in a small village in Côte d'Ivoire (Ivory Coast), West Africa, as an example. They make jewelry and other items by using a technique that has been in existence for thousands of years. It is called "lost wax casting."

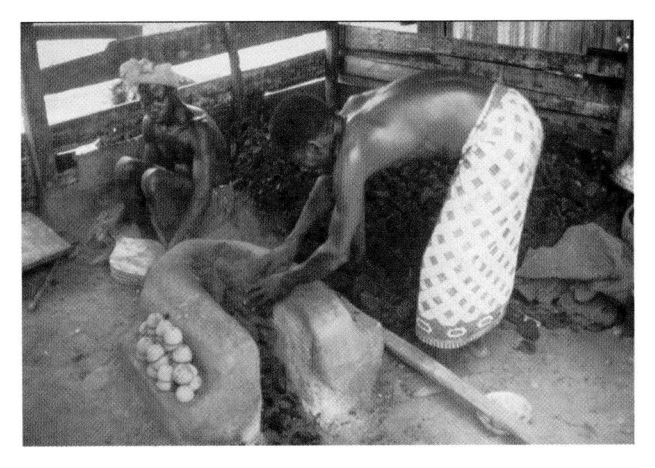

AIR IS PUMPED BY HAND BELLOWS INTO A CHARCOAL FURNACE. THIS PRODUCES THE HIGH TEMPERATURE NEEDED TO MELT GOLD.

PHOTO: Eliot Elisofon, Photographic Archives, National Museum of African Art, Smithsonian Institution, EEPA 6974

READING SELECTION

EXTENDING YOUR KNOWLEDGE

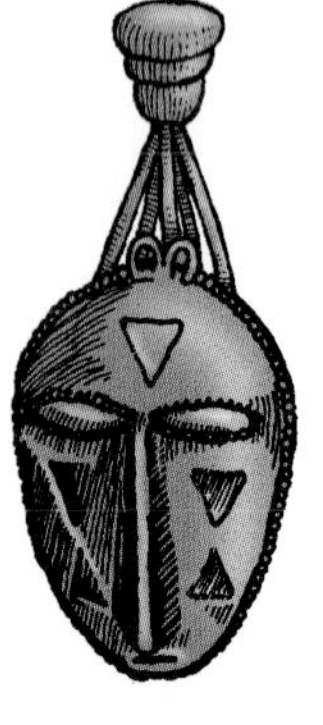

1. THE ARTIST CARVES A MODEL IN WAX (USUALLY BEESWAX). WAX IS SOFT, AND THE ARTIST CAN USE IT TO PRODUCE INTRICATE CARVINGS. AFTER COMPLETING THE MODEL, HE ATTACHES TINY WAX RODS (CALLED SPRUES) TO IT THAT WILL PRODUCE CHANNELS IN THE MOLD FOR DRAINING THE WAX AND RECEIVING THE GOLD.

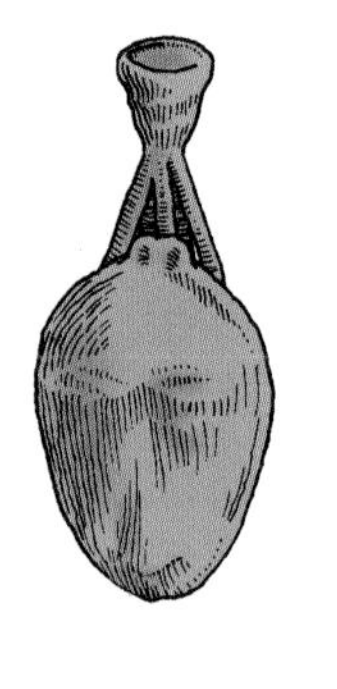
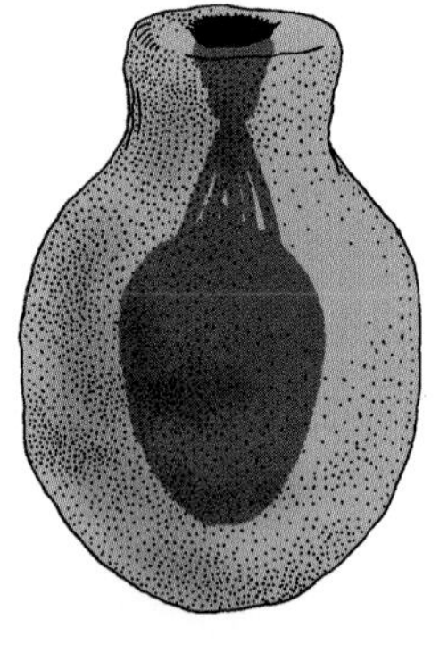

2. THE ARTIST COVERS THE WAX FIGURE WITH SEVERAL LAYERS OF FINE WET CLAY. COARSE CLAY IS THEN ADDED IN LAYERS TO COMPLETE THE MOLD. THE CLAY MOLD IS PLACED IN AN OVEN AND HEATED UNTIL IT HARDENS. THE WAX MELTS AND RUNS OUT OF THE MOLD (IN OTHER WORDS, IT'S LOST!).

3. PIECES OF GOLD ARE PLACED IN A CRUCIBLE.

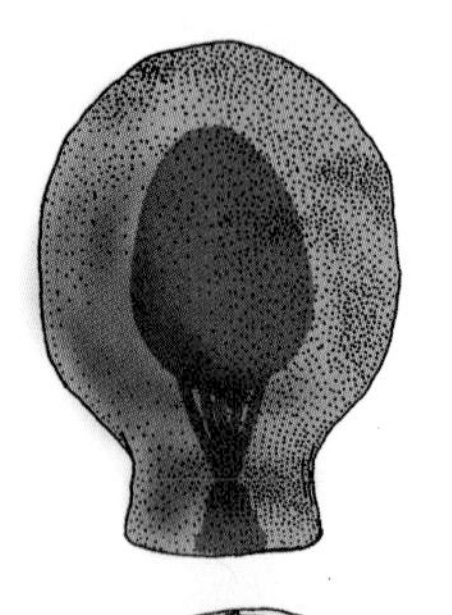
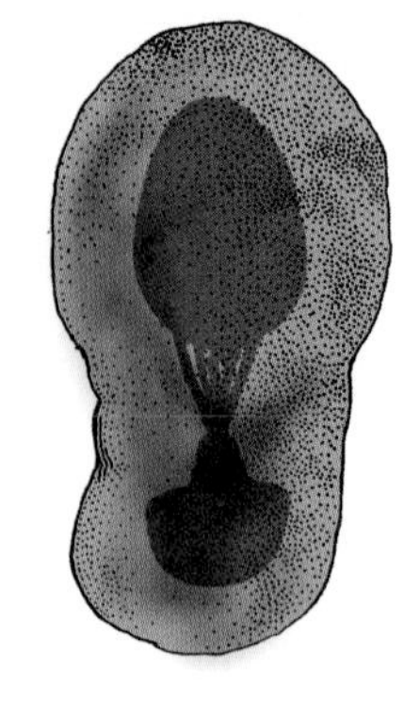

4. THE CRUCIBLE IS ATTACHED TO THE MOLD. THE TWO PARTS ARE THEN SEALED TOGETHER USING MORE CLAY.

5. THE MOLD AND GOLD ARE HEATED TOGETHER IN THE FURNACE. WHEN THE GOLD HAS MELTED, THE MOLD IS TURNED OVER SO THAT THE METAL FLOWS INTO IT. THE MOLD COOLS. THE CLAY MOLD IS CRACKED OFF, LEAVING A CASTING.

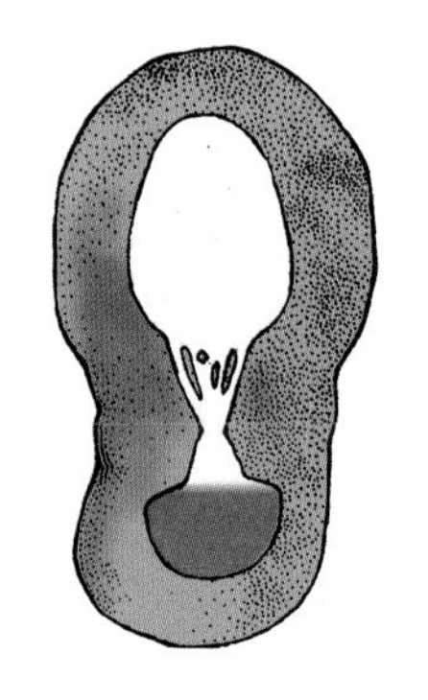
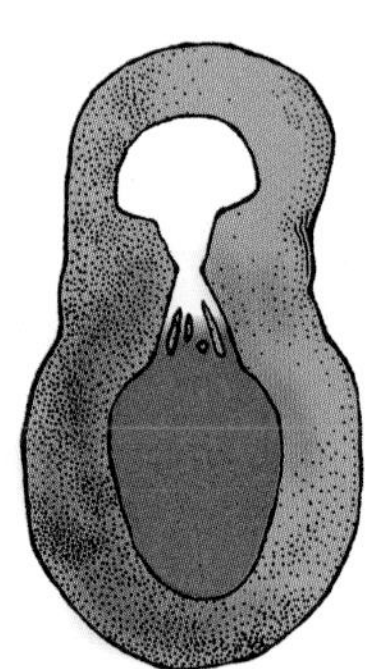

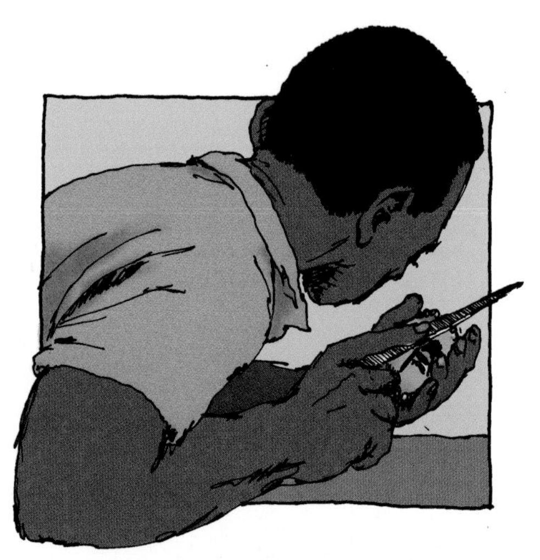

6. THE ARTIST FILES AWAY A FEW ROUGH EDGES, AND THE MASK IS READY.

In this technique, an artist produces a clay mold around an easily carved substance: wax. When the mold is heated, the wax, which has a melting point less than 70°C (158°F), melts away. The hard clay mold can then be used to produce jewelry made from metals with high melting points. The artists often use gold, which has a melting point of more than 1000°C (1832°F). The pictures at left show the major steps in lost wax casting. In these pictures, the artist is using gold.

Even modern processes make use of this ancient technique. For example, lost wax casting is sometimes still used to make small parts for machines. In many cases, the technique has been adapted using modern materials. Wax is not the only material that can be used—any material that can be burned or melted out of the mold will work. For example, engine blocks are sometimes made using a foam-casting technique. The foam is carved into the desired shape and embedded in sand. When the hot metal (traditionally iron, because it is affordable) is poured in, the foam is "lost" through evaporation, leaving the cast metal. Modern engine blocks may be made of more-lightweight materials, such as aluminum and magnesium. But, care must be taken to use strong, sturdy materials. An engine block is a solid case that seals all the combustion parts of an engine inside. If it breaks, the engine will fail! ■

THE LOST WAX METHOD HAS BEEN USED TO PRODUCE A WIDE VARIETY OF OBJECTS, INCLUDING THIS FIGURE OF A KING FROM NIGERIA.

PHOTO: Franko Khoury, National Museum of African Art, Smithsonian Institution, 85-19-12

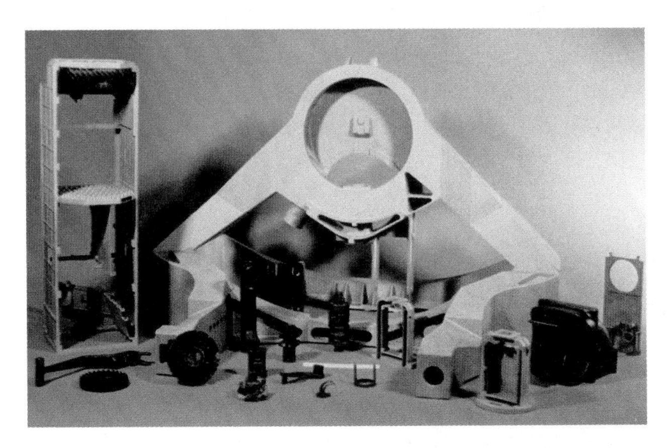

RECENTLY, NEW USES HAVE BEEN FOUND FOR LOST WAX CASTING. IT IS ONE TECHNIQUE USED TO PRODUCE PRECISION PARTS, SUCH AS THESE—DESIGNED USING COMPUTERS—FOR AIRCRAFT AND OTHER MACHINES.

PHOTO: Investment Casting Institute

DISCUSSION QUESTIONS

1. Bronze is one of the metals (alloys) often cast by artists to make large sculptures. Use library and Internet resources to find out where the largest sculptures made from such materials can be found.

2. Knowledge of melting points is important for goldsmiths. In what ways do people apply their knowledge of melting points in other occupations?

READING SELECTION

EXTENDING YOUR KNOWLEDGE

The Water Cycle: From the Sky to the Land and Back Again

THE WATER MOLECULES IN THIS LARGE, DARK CLOUD WON'T REMAIN SUSPENDED IN THE AIR FOR LONG. THIS CLOUD IS KNOWN AS A THUNDERHEAD BECAUSE IT HOLDS AN ELECTRICAL STORM. SOON, RAIN WILL POUR FROM THIS CLOUD, ACCOMPANIED BY LIGHTNING AND THUNDER.

PHOTO: Randolph Femmer/National Biological Information Infrastructure

Life as we know it could not exist without water. People need it. Animals need it. Plants need it. There's a limited supply of it.

Think about the river that flows through your town. Unless you live in an area that has a very dry climate, the river usually has plenty of water—even though the water moves in only one direction. Why doesn't the river run dry?

Water moves through the environment in a cycle. That is, it's always changing form—freezing, melting, evaporating, condensing, falling back to earth as precipitation—over and over again. The water cycle is Nature's perfect recycling project.

Water may be found in three forms or phases during this cycle. Ice is the solid phase, water the liquid phase, and water vapor the gas phase. The particles that make up water do not change from one phase to the next. The properties of each phase depend upon the temperature of the water. As the water particles get hotter they move faster and are less strongly attached to each other. This explains why a block of ice has its own shape. The water particles in the ice are strongly attracted to each other, and they hold the ice in shape.

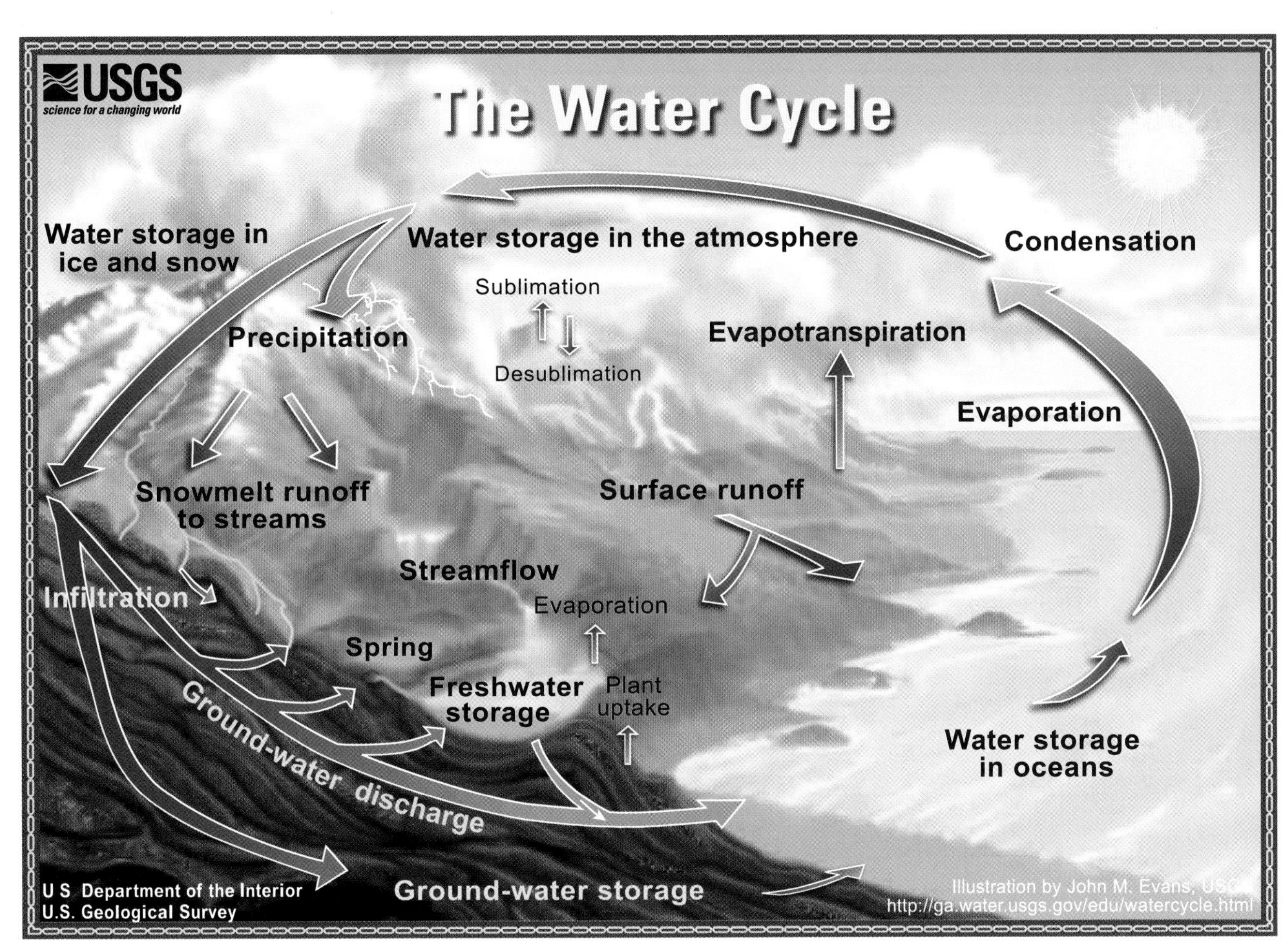

THE WATER CYCLE—FROM THE SKY TO THE LAND AND BACK AGAIN.

PHOTO: John M. Evans, U.S. Geological Survey, Colorado District

In liquid water, the water particles are not held together as strongly, which is why liquids, including water, change shape—like a drop of water from a tap—or take on the shape of a container into which they are poured. The particles in water vapor are joined by the weakest bonds. They spread out quickly, which is why matter in the gas phase has no shape.

READING SELECTION

EXTENDING YOUR KNOWLEDGE

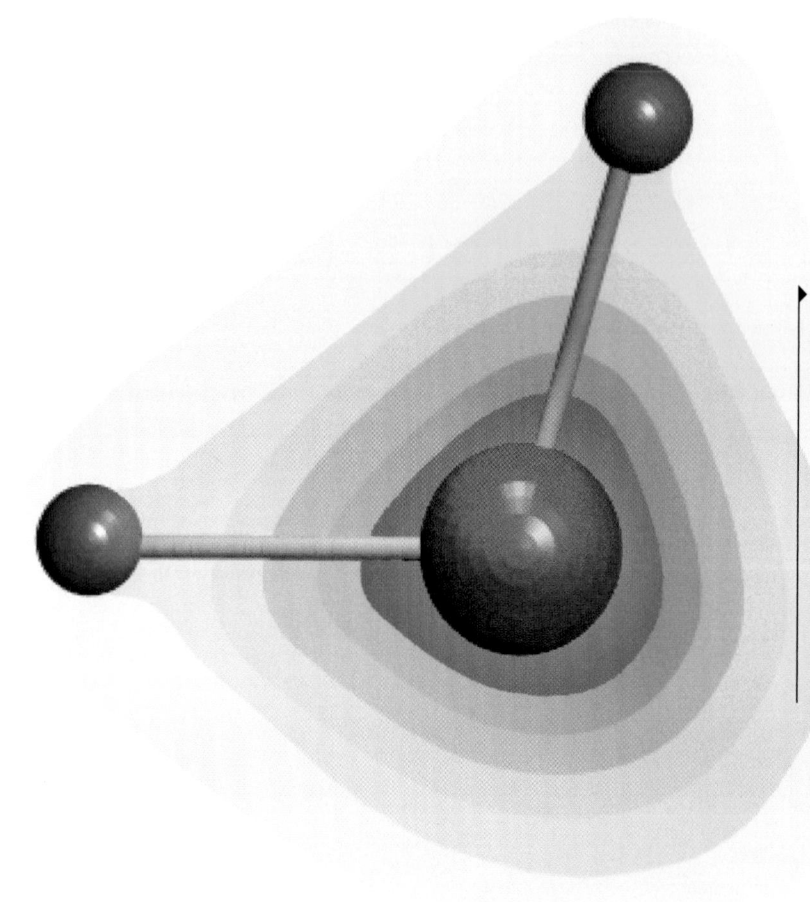

THE SMALLEST AMOUNT OF WATER IS ONE WATER MOLECULE. THIS MODEL OF A WATER MOLECULE SHOWS THAT EACH MOLECULE IS MADE UP OF TWO ATOMS OF HYDROGEN (CHEMICAL SYMBOL: H) AND ONE ATOM OF OXYGEN (CHEMICAL SYMBOL: O). THUS, THE CHEMICAL SYMBOL FOR WATER IS H_2O. INDIVIDUAL WATER MOLECULES MERGE TO FORM THE SMALL DROPLETS THAT MAKE UP CLOUDS OR FOG. THE SMALL DROPLETS MERGE TO FORM THE LARGER DROPLETS THAT BECOME RAIN, SLEET, OR SNOW.

PHOTO: Courtesy of Lawrence Livermore National Laboratory

Water is present in air as water vapor. The water particles—also called molecules—in water vapor are separate and invisible. If water vapor is invisible, why can we see clouds? You may find it surprising to learn that clouds are made up of liquids or tiny particles of ice, and not water vapor. When water vapor is cooled, the water molecules get closer together in a process called condensation and eventually form tiny droplets of water. These droplets combine with others to form larger droplets. As long as these droplets stay small they remain suspended—held up—in the air as clouds or, at ground level, as fog.

As these droplets get bigger they become heavier, and they may eventually fall from the clouds as rain, or, under very cold conditions, as ice in the form of sleet, hail, or snow. This process is called precipitation. Once the precipitation reaches the ground-level, it can take many paths:

- It can fall directly onto streams, lakes, or oceans.
- It can fall on land.
- It may soak into the soil. Once in the soil, it may be taken up by plants, flow beneath the soil surface to eventually join surface water, or seep into the spaces in the rock that lies beneath the soil.
- It may be used by people for drinking, washing, preparing food, and many other purposes.

Most of the water eventually returns back to the air through evaporation—turning from liquid water to water vapor. One special type of evaporation is called transpiration. Transpiration is evaporation from the surface of plants. Plants take up water through their roots and then release water through their leaves. The air near the ground, which contains lots of water vapor, rises as it is pushed up by cooler, more dense air moving downward. As the air rises it expands and cools. The water vapor in the cooling air condenses. It once again forms clouds. The water cycle is complete and starts all over again. ■

DISCUSSION QUESTIONS

1. What human activities interrupt the water cycle?
2. What natural resources, other than water, cycle through earth's ecosystems?

LESSON 8

CHANGING MATTER AND MASS

A GLACIER IN ALASKA CALVES A NEW ICEBERG. THE ICEBERG WEIGHS THOUSANDS OF TONS. WHAT HAPPENS TO THE MASS OF ITS MATTER WHEN IT MELTS?

PHOTO: Commander John Bortniak, NOAA Corps

INTRODUCTION

In Glacier Bay, Alaska, 12 giant glaciers meet the sea. As the seawater undermines and melts these huge rivers of ice, icebergs break off, or "calve." Chunks of ice, some more than 60 meters (about 200 feet) high and weighing thousands of tons, plunge with a huge splash into the sea. Because ice is less dense than water, the icebergs float. For a few weeks in the summer, they may provide floating islands for the local wildlife before becoming part of the sea themselves. What happens to the mass of a 50,000-ton iceberg (that's more than 100 million pounds) when it melts? For that matter, what happens to the mass of a melting ice cube in a glass of soda, the mass of water as it freezes in the freezer, or the mass of boiling water in a kettle when it turns into steam?

What happens to the mass of matter when it changes phase? In this lesson, you will discuss and try to answer this question.

OBJECTIVES FOR THIS LESSON

- Discuss what happens to the mass of a substance when it changes state.
- Conduct an experiment to investigate whether a change in mass occurs when ice melts.
- Discuss sources of experimental error within your experiment.
- Design an inquiry to test your own prediction about a change in mass that may occur when water freezes.

MATERIALS FOR LESSON 8

For you

1 copy of Student Sheet 8.1: Investigating Mass and Melting
1 copy of Student Sheet 8.2: Investigating Mass and Freezing

For you and your lab partner

1 plastic soda bottle with screw cap
1 250-mL beaker
1 paper towel
2 ice cubes, crushed
Access to an electronic balance

GETTING STARTED

1. What happens to the mass of matter when it changes phase? Write in your science notebook your prediction of what will happen to the mass of the matter in the following situations:

   A. the mass of an ice cube when it melts

   B. the mass of water in an ice cube tray when it freezes

   C. the mass of the water in a tea kettle when it boils

2. Your teacher will record your predictions. Be prepared to contribute your predictions for each phase change.

AT THIS HOT SPRING IN YELLOWSTONE PARK, WHAT HAPPENS TO THE MASS OF WATER WHEN IT EVAPORATES? DOES THE WATER VAPOR HAVE THE SAME MASS AS THE LIQUID WATER THAT FORMED IT?

PHOTO: © David Marsland

INQUIRY 8.1

INVESTIGATING MASS AND MELTING

PROCEDURE

1. One member of your group should obtain the plastic box containing the apparatus. Divide the apparatus equally between each pair in your group.

2. In this inquiry, you will work with your partner to test your prediction about what will happen to the mass of ice when it melts. Examine your apparatus carefully. Discuss with your partner how you could use this apparatus, crushed ice, and an electronic balance to test your prediction.

3. Under Step 1 on Student Sheet 8.1: Investigating Mass and Melting, write the procedure you will use to investigate what happens to the mass of ice when it melts.

4. Your teacher will ask some pairs of students to share their procedures with the whole class. Be prepared to contribute your ideas to the discussion.

5. During the discussion, the class will agree on a procedure for the experiment. Write the class procedure under Step 2 on Student Sheet 8.1.

6. Design a results table under Step 3.

7. Begin the procedure. Record all of your results in your results table.

8. While the ice is melting, complete Steps 1 and 2 in Inquiry 8.2.

9. When the ice has melted, measure the mass of the bottle and water (as outlined in the class procedure) and record the mass of the apparatus in your results table. Don't forget to wipe any condensation off the outside of the beaker with a dry paper towel.

10. Write a description of what happened to the mass of the ice when it melted. Why did you use a sealed container for this experiment? Write an explanation.

11. You will be asked to contribute your results to a class results table. Use the data from all the pairs to complete Table 1 on Student Sheet 8.1.

12. Did all of the pairs obtain the same result? Complete Step 6 on Student Sheet 8.1.

13. Participate in a class discussion on experimental error.

14. Compare the measurements you entered in Table 1. Use them to complete Steps 7a through 7d on Student Sheet 8.1.

INQUIRY 8.2

INVESTIGATING MASS AND FREEZING

PROCEDURE

1. Do you think any change in mass occurs when water freezes? Discuss your ideas with your partner.

2. How could you find out whether any change in mass occurs when water freezes? Design an experiment to answer this question. Describe your ideas for a procedure under Step 1 on Student Sheet 8.2: Investigating Mass and Freezing.

3. Your teacher will ask you to describe your procedure. After a discussion, one student in your class will do the experiment following the class procedure. Record the mass of the sealed bottle of water used in the experiment.

4. You will revisit this experiment in a later lesson. When you have the results of this experiment, complete Step 3 on Student Sheet 8.2.

REFLECTING ON WHAT YOU'VE DONE

1. Participate in a class discussion on the conservation of mass.

2. Write a paragraph in your science notebook explaining how conservation of mass applies to melting.

Choosing Materials for Pedal-Powered Flight

DAEDALUS MADE WINGS FROM WAX AND FEATHERS TO ESCAPE KING MINOS.

In Greek mythology, Daedalus made wings out of wax and feathers. Daedalus used the wings to escape from the prison of King Minos on the Greek island of Crete. He flew more than 100 kilometers (62.1 miles) to safety.

In the mid-1980s, airplane engineer John Langford decided to do in real life what Daedalus had done in the myth. However, Langford didn't use wax and feather wings. Instead, he put together a team that built an ultra-lightweight, pedal-powered airplane. He called the plane *Daedalus*, after the mythological Daedalus.

Kanellos Kanellopoulus, a Greek Olympic bicyclist, was the pilot. He flew the plane 115 kilometers (71.5 miles) from Crete to another Greek island, Santorini. As of 2009, the plane still held the distance record for human-powered flight.

It is easy to see where the myth of Daedalus came from. Almost everyone, at sometime or another, dreams of flying like a bird. Through the centuries, many would-be eagles have strapped on homemade wings and tried to fly using only their own muscles. They flew like bricks.

The problem, says Langford, is that people are not very good engines. To fly the *Daedalus* 115 kilometers would take several hours. For that length of time, even the best athletes can sustain only the same power output as a bright lightbulb. Because of this lack of power, the plane had to be very light. "Every gram mattered," says Langford.

READING SELECTION

EXTENDING YOUR KNOWLEDGE

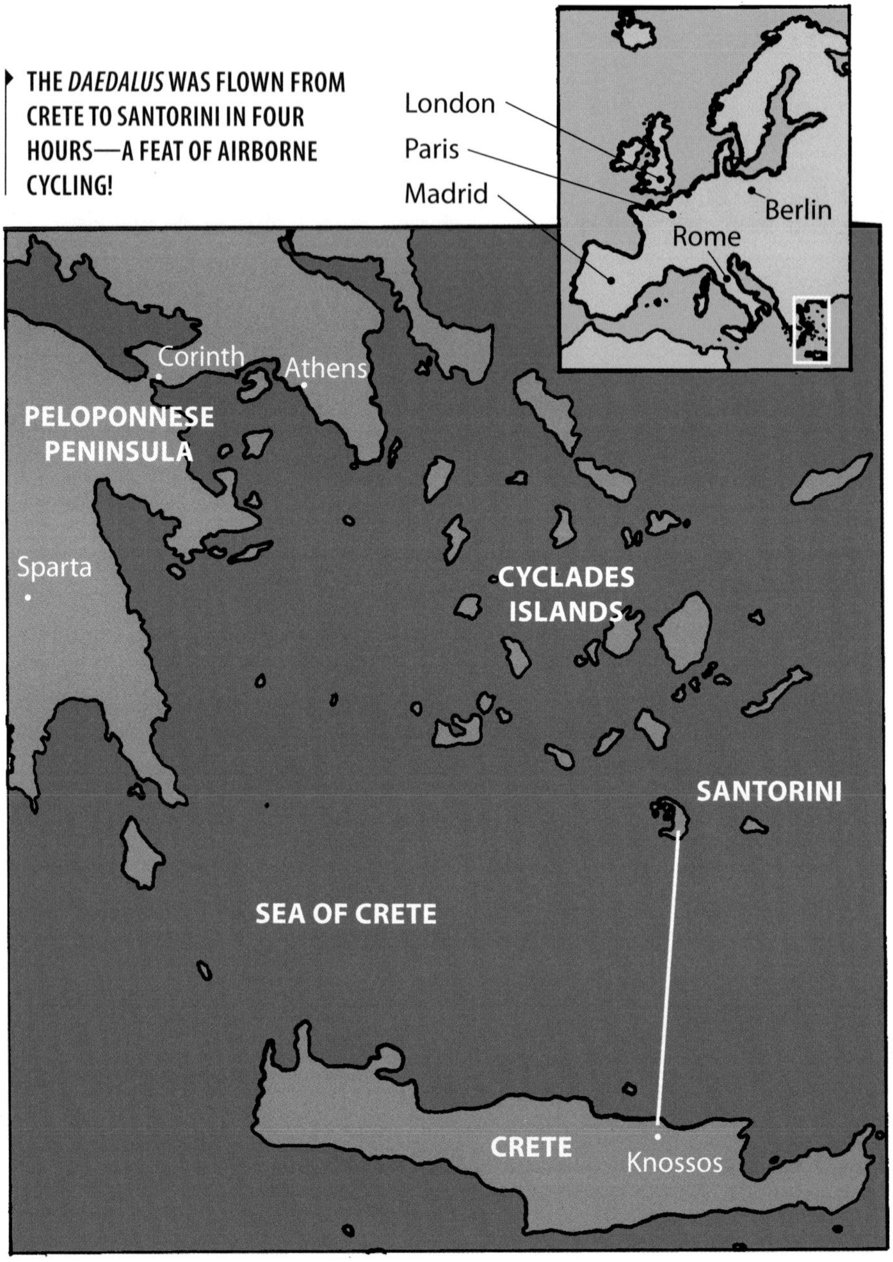

THE *DAEDALUS* WAS FLOWN FROM CRETE TO SANTORINI IN FOUR HOURS—A FEAT OF AIRBORNE CYCLING!

When Langford and his team set out to build *Daedalus*, they needed materials that were very strong and very light. Fortunately, new materials had been developed that fit their needs.

For the plane's frame, Langford chose tubes made of a carbon composite. This material is composed of thin, very strong carbon threads embedded in plastic. The wings were made of solid foam building insulation cut to the proper shape. The plane's skin was made of Mylar™, an extremely thin but sturdy sheet material that is also used in videotapes and shiny helium balloons.

These materials are said to be "lightweight." But one pound of the lightest carbon composite still weighs as much as one pound of steel. The materials used to build *Daedalus* were special because of their combination of high strength and low density. This combination allows these materials to pack a lot of strength into relatively little weight.

Even though they used the most advanced materials, the builders of the plane couldn't make it any stronger. To make it stronger, they would have had to use denser materials, which would have made the plane too heavy to fly the distance. Despite a wing span of 34 meters (112 feet)—almost as wide as a passenger jet—the plane weighed only 31 kilograms (68.3 pounds). It was so fragile that the flight could only succeed on a completely windless day. But succeed it did.

At 7:00 a.m. on April 23, 1988, *Daedalus* took off from Crete and arrived at Santorini about four hours later. Once there, the plane's fragility caught up with it. While being maneuvered to land on the beach, it was hit by a gust of wind. The plane broke into pieces, and pilot Kanellopoulus cooled off with a short unscheduled swim.

Langford still dreams of having another try at human-powered flight. "It would be great fun to try to make a plane using only the materials the ancient Greeks had," says Langford. It would be a tough job, though. "I think I could make one that would fly, using thin silk for the skin and bamboo for the skeleton," he says. "But without modern lightweight materials, you could never go as far as *Daedalus*." ■

MANY EARLY ATTEMPTS AT HUMAN-POWERED FLIGHT, LIKE THAT MADE IN THIS ORNITHOPTER, FAILED BECAUSE OF POOR DESIGN AND THE LACK OF STRONG, LOW-DENSITY MATERIALS.

PHOTO: Library of Congress, Prints & Photographs Division, LC-B2-943-5

THE *DAEDALUS* PEDAL-POWERED PLANE FOLLOWS THE ROUTE OF THE DAEDALUS OF MYTHOLOGY.

PHOTO: NASA Dryden Flight Research Center

DISCUSSION QUESTIONS

1. What two properties did the designer of the *Daedalus* look for in the materials he would use to build a human-powered airplane? Why were these important?
2. Imagine you have been asked to design each of the following items:
   A. Raincoat
   B. Bullet-proof vest
   C. Milk bottle
   D. Fishing weight (sinker)

   What three properties would you look for when selecting materials to make each item? What materials would you use to make each?

LESSON 9

EXPLORATION ACTIVITY: A MANUFACTURED OBJECT

INTRODUCTION

In this lesson, you will begin a research activity that you will work on over the next several weeks. This project will give you the opportunity to apply what you have learned in the unit to the world around you. In this research activity, you and a partner will select a simple manufactured object. Outside of class, you will investigate the chemistry, technology, and history of the object by doing research at the library and on the Internet. You will then compile the information you have collected to create an exhibit. An oral presentation on one of the materials that makes up the object you choose will conclude the project. The work you do for the Exploration Activity will be an important part of your grade for this unit.

WHAT MATERIALS ARE CHOSEN TO MAKE THIS SKATEBOARD? HOW ARE THESE CHOICES MADE?

PHOTO: U.S. Air Force Courtesy Photo

OBJECTIVES FOR THIS LESSON

- Select a simple manufactured object to research.
- Conduct library and Internet research on the major materials that make up the object you have chosen.
- Create an exhibit based on your research.
- Give an oral presentation on one of the materials that makes up the object you have chosen.

MATERIALS FOR LESSON 9

For you

1 copy of Inquiry Master 9a: Scoring Rubric for the Cube
1 copy of Inquiry Master 9b: Scoring Rubric for the Oral Presentation
1 copy of Student Sheet 9a: What Are Bikes Made From and Why?
1 copy of Student Sheet 9b: Exploration Activity Schedule

For you and your partner

Clear tape or glue
Scissors
Card stock, poster board, or lightweight cardboard

GETTING STARTED

1. Contribute to a class discussion on the reading selection from Lesson 8, “Choosing Materials for Pedal-Powered Flight,” and the accompanying questions.

2. Read “The Right Material.”

3. One of your classmates may have brought a bicycle to class. It is an example of a manufactured object. Discuss the choice of materials that make up the bicycle. After the class discussion, work with your group to complete Table 1 on Student Sheet 9a: What Are Bikes Made From and Why?. List the function, the type of material, and the properties of the material for each bicycle part.

INTRODUCING THE EXPLORATION ACTIVITY

PROCEDURE

1. After your teacher gives you Student Sheet 9b: Exploration Activity Schedule, tape it to the inside front cover of your science notebook. You will need to refer to it as you work on the Exploration Activity. Follow it carefully, or you may lose points.

2. Follow along as your teacher reviews the Exploration Activity Guidelines.

READING SELECTION

BUILDING YOUR UNDERSTANDING

THE RIGHT MATERIAL

Matter is used to make things. The term "technology" refers to the way people alter and shape matter so that it can be used to make things. For example, gold can be found as metal nuggets. But through technology (for example, the lost wax method), it can be fashioned into jewelry. Through different technology, it can be used to place electronic components inside a computer.

Part of making any useful object is choosing the right kind of material for it. Materials can be any type of matter, from the metal in a bicycle frame to the compressed air in a bicycle tire. Some materials are used directly from nature (for example, wood and stone). These are called raw materials. Other materials are made from raw materials that are refined or processed in some way. For example, one of the raw materials used to make glass is sand.

Think about some everyday manufactured objects. Some are very complex. For example, a car contains thousands of different parts and is made from hundreds of different types of matter. Each material used to build the car is chosen for the job it must do. That job is its function. How is this choice made? It is based on several factors, including cost, availability, and, most importantly, the properties of the material.

Scientists and engineers are often on the lookout for better, cheaper, or more readily available materials to replace the traditional materials used in objects. They try to find or design materials that have the right properties. For example, most shoes were once made entirely from pieces of leather that were sewn or nailed together. Nowadays, many shoes are made from a variety of materials, each suited or designed to fit the function of that part of the shoe. Soles may consist of a combination of durable, shock-absorbing rubbers or plastics. Uppers are often made of waterproof, breathable synthetic fabrics or stain-resistant plastics with soft linings that cushion the foot and protect it from abrasion.

WHAT PROPERTIES MADE ALUMINUM AND TRANSPARENT PLASTIC IDEAL MATERIALS FOR THE MANUFACTURING OF THIS DISK?

PHOTO: The Artifex/creativecommons.org

During the Exploration Activity, you will learn more about the materials that make up an object. You will study the relationship between object function, the choice of materials to make the object, and the properties and origins of the materials. ■

EXPLORATION ACTIVITY GUIDELINES

PART 1

CHOOSING THE OBJECT

PROCEDURE

1. To make your work easier, you and your partner should choose a relatively simple manufactured object. Discuss with your partner which object you are going to research. Some examples are given in the list titled "Exploration Activity Objects." You can choose one of these or think of another that you use every day. But remember, keep the object simple. An object made from two or three materials will be much easier to research and present than one made from many materials. You do not need to make a final decision immediately, but your teacher will set a deadline.

EXPLORATION ACTIVITY OBJECTS

Aerosol can	Ballpoint pen
Battery	Bottle
CD/DVD	Clothing
Cooking pot	Diaper
Felt-tip pen	Football
Furniture	Golf ball
In-line skates	Lightbulb
Magnifier	Matches
Notebook	Pencil
Pencil sharpener	Scissors
Sneakers	Soda can
Tape dispenser	Thermometer
Tools	Toothbrush
Toy	

2. During the next week, meet with your partner and write a short paragraph identifying the object you have chosen and your reasons for selecting it (Figure 9.1 shows a sample paragraph). Hand it in by the due date on your schedule. Your teacher must approve your object. If too many pairs choose the same object, you may be asked to select another.

Jane Lionel and Juanita Lopez

We have chosen eyeglasses for our piece of research. Both of us wear glasses for reading. We are going to take Juanita's glasses as our example.
They are made from three types of substanses. Some type of golden metal, they have glass lenses and plastic nose pieces as well as plastic pads on the stems.

SAMPLE PARAGRAPH IDENTIFYING AN OBJECT AND OUTLINING THE REASON IT WAS CHOSEN
FIGURE **9.1**

PART 2
STARTING THE RESEARCH
PROCEDURE

1. Start gathering information about your object. Your information will be divided into five sections. As you gather information, write your notes under these headings:

FUNCTION Explain what the object does or what it is used for.

MAJOR MATERIALS List the main materials from which the object is made.

WHY THESE MATERIALS WERE CHOSEN Tell what properties of the materials make them good choices for use in the object.

ORIGIN OF ONE OF THE MATERIALS Select one major material in the object and investigate its raw materials, where they are found, and the processes they undergo to become usable in the object.

HISTORY OF THE OBJECT Was it invented? If so, by whom? When and where did it first appear? How do the original designs and choice of materials differ from those in use today?

Exploration Activity Part 2 continued

2. The section "Origin of One of the Materials" is similar to (but not as detailed as) the topic of your oral presentation. See Part 4: Giving the Oral Presentation (on page 119) to find out what you should research for your presentation.

3. Use your notes to help you conduct a brainstorming session with your partner. After the brainstorming session, write an outline of your investigation. The outline should be in a format similar to that shown in Figure 9.2.

EXAMPLE OF EXPLORATION ACTIVITY RESEARCH
FIGURE **9.2**

4. On another sheet of paper, prepare your bibliography. The bibliography can include books, newspapers, magazines, and TV programs. You should have at least one Internet and one CD-ROM or DVD reference (see Figure 9.3).

5. Hand in your outline and bibliography on or before the due date on your schedule. Your teacher will use this information to make sure your research is heading in the right direction.

Jane and Juanita

These are the references we have found so far.

The Physical World by Martin Sherwood and Christine Sutton, Oxford University Press (from Bethesda Library)

Chemistry by Ann Newmark, Eyewitness Books, Dorling Kindersley (from the school library).

Grolier Multimedia Encyclopedia (on CD ROM) (We borrowed it from Juanita's sister)

Asimov's Chronology of Science and Discovery by Isaac Asimov, Harper and Row (Bethesda Library)

We also took the glasses to the optician in Bethesda to find out the type of metal the frames are made from.

We need some help finding out where plastics come from.

EXAMPLE OF A BIBLIOGRAPHY
FIGURE **9.3**

PART 3

CREATING THE CUBE

PROCEDURE

1. Continue your research. At least a week before the due date for the completed exhibit, gather together all of your information.

2. Inquiry Master 9a: Scoring Rubric for the Cube explains what is required for each section of your exhibit. Review it carefully because it tells you how you can earn high scores for your exhibit. Notice that points are awarded for the bibliography and for the presentation, including the use of an effective design for your exhibit. Remember, points will be deducted for late work.

3. Write each section of your exhibit (use the five headings listed in Part 2, Step 1 of the Exploration Activity Guidelines). If you can, use a word processor to type the final text. You have very limited space for each of the sections. Choose the content, including pictures and diagrams, very carefully.

4. Construct your exhibit. Follow the instructions in Steps 5 and 6 to make the cube.

5 Make the cube from lightweight cardboard, card stock, or poster board. The dimensions of the cube should be about 15 × 15 × 15 cm. Figures 9.4 and 9.5 show how to assemble the cube.

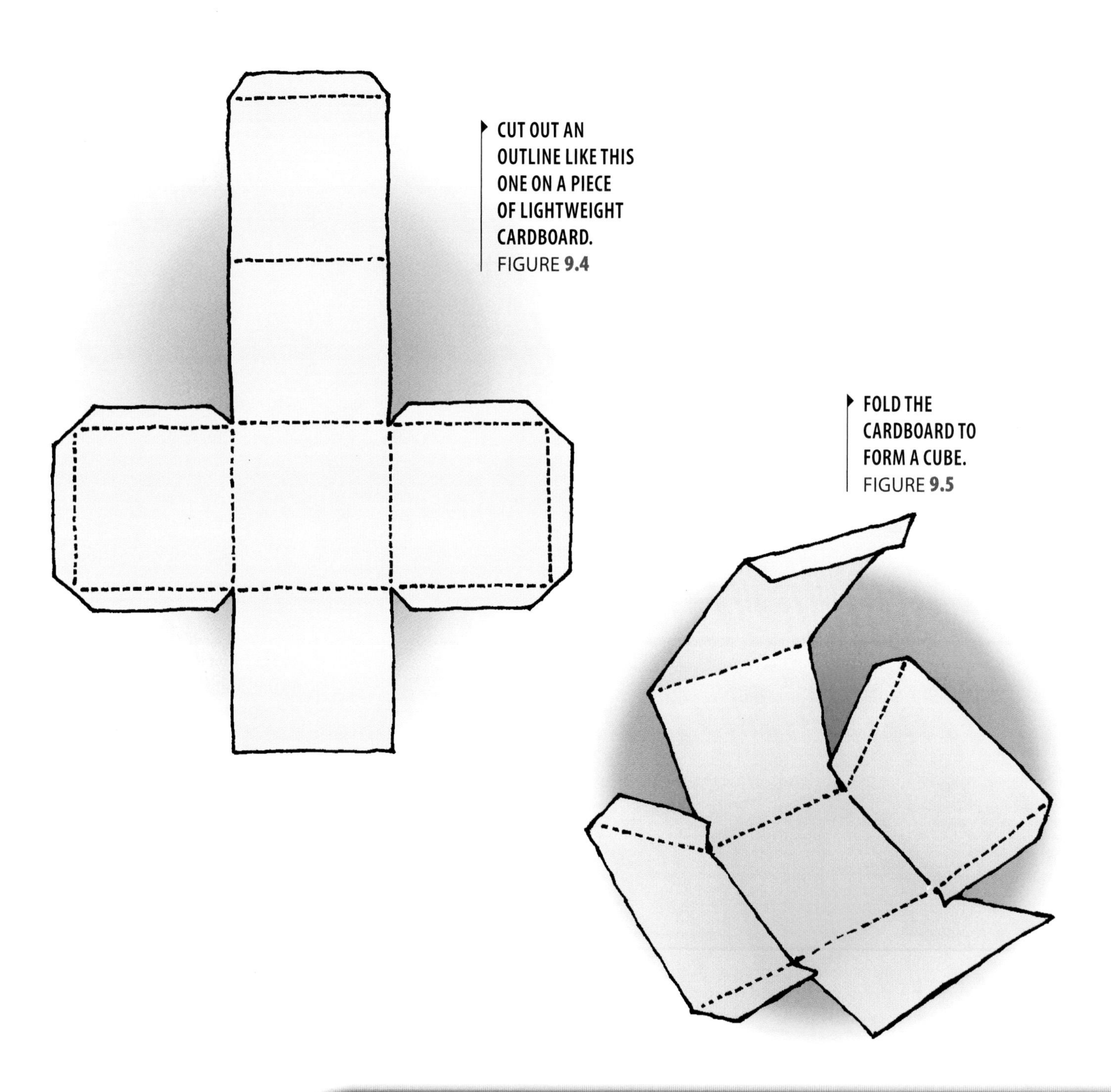

CUT OUT AN OUTLINE LIKE THIS ONE ON A PIECE OF LIGHTWEIGHT CARDBOARD.
FIGURE **9.4**

FOLD THE CARDBOARD TO FORM A CUBE.
FIGURE **9.5**

Exploration Activity Part 3 continued

6. Use one side of the cube for each heading (this will use five sides). Make sure that at least four sides of the cube read in the same direction (see Figure 9.6). Put a picture or photograph (or, if possible, the object) on the sixth side. It is important that your names and bibliography appear somewhere on the exhibit. Figures 9.7 and 9.8 show completed cubes.

7. If your cube is too small to hold all your information, include the additional information in your oral presentation.

8. Hand in your exhibit on or before the due date.

AT LEAST FOUR SIDES OF THE CUBE SHOULD READ IN THE SAME DIRECTION.
FIGURE **9.6**

THE STUDENTS WHO BUILT THESE CUBES HAD FUN RESEARCHING AND DESIGNING THEM.
FIGURE **9.7**
PHOTO: ©David Marsland

ONCE COMPLETE, THE CUBES FROM YOUR CLASS WILL MAKE AN EXCITING EXHIBIT ABOUT MATERIALS AND HOW WE USE THEM. ALTHOUGH EACH CUBE FOLLOWS THE SAME FORMAT, YOUR PERSONAL TOUCHES WILL MAKE YOUR CUBE UNIQUE.
FIGURE **9.8**
PHOTO: ©David Marsland

PART 4

GIVING THE ORAL PRESENTATION

PROCEDURE

1. Work with your partner to prepare a short oral presentation. It should focus on the origin of one of the materials that make up your object. You should provide detailed information on the following topics:
   - One of the materials from which your object is made
   - The properties of the material
   - The properties of the material that make it a good choice for use in your object
   - One of the raw materials from which the material is made
   - The geographical source or sources of the raw material
   - How the raw material is extracted and/or processed before it is used in your object

2. Both you and your partner should be involved in giving the presentation. During your presentation, use some visual aids such as posters, maps, computer presentations, or overhead transparencies. If you can, use Web pages or a short video.

3. Carefully read Inquiry Master 9b: Scoring Rubric for the Oral Presentation. It tells you how your oral presentation will be assessed. Use the table to plan your presentation.

4. With your partner, practice giving the presentation. Time yourselves so that the presentation is between 3 and 5 minutes long.

5. Make sure you have all of your materials ready before you give your presentation. You may refer to notes during your presentation, but you should avoid reading them.

WHEN WILL YOU DO ALL OF THIS WORK?

You will be given several homework assignments and a small amount of class time to do this work. However, you will have to do most of it on your own time. At the end of the unit, two to three class periods will be used for oral presentations.

BICYCLE INGREDIENTS

Compared to cars, bicycles look pretty simple. But this appearance is deceiving. Even an inexpensive bike can be made of more than one hundred different materials. These include several kinds of steel, other metals such as chrome and aluminum, several kinds of rubber, a few oils, and different types of plastic.

Dan Connors is an engineer with Cannondale Corporation, a company in Connecticut that makes bikes. He says choosing the material for each bicycle part always boils down to a trade-off between strength, weight, and price. "You want all the parts to be strong, but they can't weigh too much. Nobody wants to pedal around with a bunch of excess weight," he says. But price is important, too. "You can design the greatest thing in the world. But if that means putting a thousand dollar part on a bike you want to sell for five hundred bucks, you won't get too far," Connors says.

The single biggest part of a bicycle is the frame. The first bike frames were made of wood. Today, most bike frames are made of steel. Steel has a good balance between strength and weight. It is easy to work with. It also doesn't cost much. For more expensive frames, designers often choose aluminum. Aluminum or aluminum alloys can have the same strength as steel, but they weigh less. Aluminum is also cheap to buy. Unfortunately, it is tricky to weld aluminum pieces together, so aluminum frames cost more.

THE DRAISIENNE, INVENTED IN 1818, WAS THE FIRST TWO-WHEELED MACHINE FOR PERSONAL TRANSPORT. IT HAD NO PEDALS AND WAS MADE FROM IRON AND WOOD, THE MOST PRACTICAL MATERIALS AVAILABLE AT THAT TIME.

BICYCLES LIKE THIS ONE, MADE PARTIALLY FROM CARBON COMPOSITES, ARE LIGHTER THAN METAL BIKES AND CAN BE DESIGNED TO BE MORE AERODYNAMIC THAN THOSE MADE FROM METALS.

PHOTO: Courtesy of Cannondale, by Tim DeFrisco

READING SELECTION

EXTENDING YOUR KNOWLEDGE

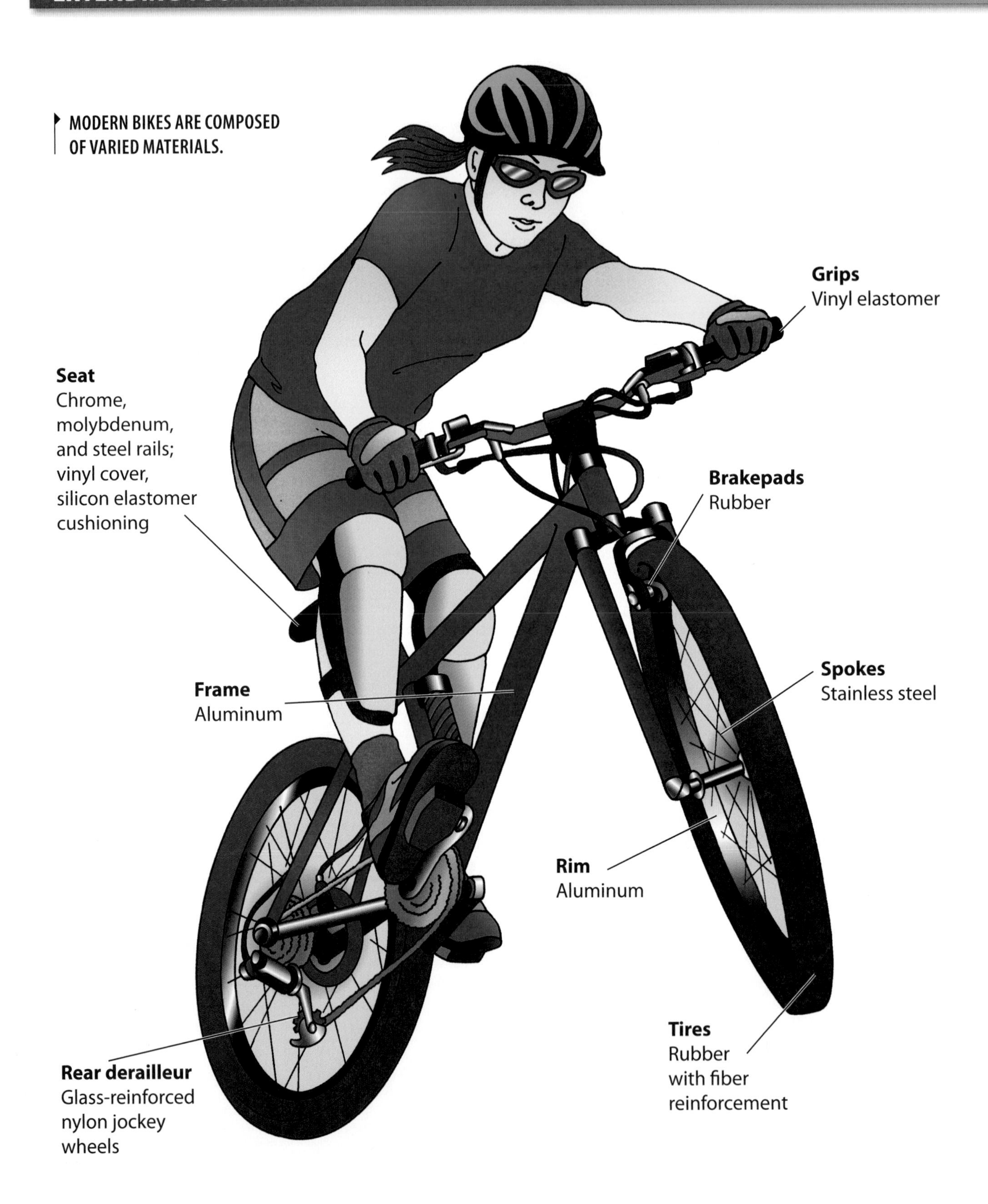

Some very high-priced bikes have frames made of carbon composites. These new materials are made by setting strong carbon fibers in a solid plastic matrix. Frames made of carbon composite can be as strong as steel but weigh only one-third as much, says Connors.

Strength, weight, and cost are important for other parts of a bike, too. Take the gears, for example. The big front gear turned by the pedals does not need to be as strong as the gears on the back wheel. Designers often save a little weight by using aluminum alloy for the front gear, says Connors, but this trick won't work at the back. "You might save a little weight if you put aluminum alloy gears in the back, but the gears would wear out after a couple of months," he says. The back gears are usually made from more durable steel.

Tires are made of rubber. Rubber is flexible and holds air, but by itself, it is not very strong. To compensate for the lack of strength of rubber, bicycle tire makers embed long fibers, often made from nylon, inside the rubber. The fibers help the tire hold its shape and resist punctures.

The ball bearings inside the axles of the wheels are especially hardened steel. They may be sealed with lubricants that have special additives to withstand heat.

You might think that it would be very difficult to improve on something that has been around for more than a hundred years, like the bicycle. Fortunately, new materials are constantly being discovered or invented. This gives engineers like Dan Connors new options for designing bicycles. ■

DISCUSSION QUESTIONS

1. If you designed a bicycle, what features would you add? What materials would you use, and why? Draw a picture of your design.

2. Research how the design and materials of a sports product work to improve performance.

LESSON 10

WHAT HAPPENS WHEN SUBSTANCES ARE MIXED WITH WATER?

WHY DOESN'T THE SAND ON THIS BEACH MIX EASILY WITH THE WATER IN THE OCEAN?

PHOTO: Randolph Femmer/National Biological Information Infrastructure

INTRODUCTION

What happens when different substances are mixed with water? Do they all behave in the same way? Does the type of mixture that a substance forms with water depend on the properties of the substance? In this lesson, you will investigate what happens when you mix several pure substances with water. Using your observations, you will identify some of the characteristics of solutions. You will also discuss the terms that are used to describe the formation of solutions.

OBJECTIVES FOR THIS LESSON

- Observe what happens when different substances are mixed with water.
- Identify the characteristics of solutions.
- Define and use some terms that describe the parts of a solution and the processes that take place when a solution is formed.

MATERIALS FOR LESSON 10

For you

1 copy of Student Sheet 10.1: Mixing Substances with Water
1 pair of safety goggles

For you and your lab partner

1 test tube rack
5 test tubes
2 rubber stoppers
Access to water

For your group

5 jars containing these substances:
Copper (II) sulfate
Sodium chloride
Zinc oxide
Sulfur
Confectioners' sugar
1 lab scoop
1 test tube brush

GETTING STARTED

1. Your group will be given a test tube containing a mixture of food coloring and water. In your science notebook, list all the properties of this mixture that you can observe.

2. The class will discuss these observations and relate them to the mixtures you will investigate in this lesson.

3. Return the tube of food coloring to your teacher. Keep the beaker; you can use it to collect water for Inquiry 10.1.

SAFETY TIPS

Wear safety goggles throughout the inquiry.

If you splash a solution in your eyes, immediately flush your eyes with a lot of water and report your accident to your teacher.

Do not mix the contents of different test tubes.

When you complete the inquiry, wash your hands.

WHAT HAPPENS TO THIS SOLID SUBSTANCE (OATS) WHEN YOU MIX IT WITH WATER?

PHOTO: S. Diddy/creativecommons.org

ADDING WATER TO SUBSTANCES

PROCEDURE

1. In Inquiry 10.1, you will work in pairs, but you will discuss your results with the other pair in your group.

2. One person from your group should obtain a plastic box containing the materials. Check the contents of the plastic box against the materials list. You will be sharing the jars containing the substances, the lab scoop, and the test tube brush with other members of your group, but make sure your pair has one set of the remaining apparatus.

3. You have samples of five different substances. You are going to investigate what happens to each when you add water to them.

4. Put one lab scoop of copper (II) sulfate into a test tube.

5. Add water to a depth of 5 cm.

6. Seal the test tube with a rubber stopper.

7. Shake the mixture 10 times. Do not knock the tube against the desk.

Inquiry 10.1 continued

8. Examine the contents of the tube (see Figure 10.1). Observe what happens to the solid substance you put in the tube. Write the name of the substance in Table 1 on Student Sheet 10.1: Mixing Substances with Water. Describe the appearance of the contents in the appropriate space in the table.

9. Repeat the procedure with the remaining four substances.

10. Discuss your results with the other members of your group. Complete the third column of Table 1. Do not clean up your remaining materials until after the class discussion. Put the zinc oxide waste into the container provided. Wash the sulfur down the drain with a lot of water. Clean your apparatus and return it to the plastic box.

LOOK AT YOUR MIXTURE. IS IT TRANSPARENT? IS IT OF UNIFORM COMPOSITION? IS IT A SOLUTION?

FIGURE **10.1**

REFLECTING
ON WHAT YOU'VE DONE

1. Discuss the results of Inquiry 10.1 with the rest of the class.

2. Observe carefully as your teacher shows you what happens when water is added to potassium permanganate. After the demonstration, write a full description of what happened on Student Sheet 10.1. Use the terms that have been discussed during the lesson. Look at the terms listed in Step 4 (below) if you are unsure what these words are.

3. Your teacher will repeat the demonstration using sand. Describe your observations as before.

4. On Student Sheet 10.1, write your definitions of the following terms: soluble, insoluble, solvent, solute, solution, and dissolve.

DISSOLVING HISTORY

Dateline: January 1998, Athens, Greece

A team of archaeologists, architects, ironworkers, and marble cutters has just started a new project. Its goal? To restore the Temple of Athena, a masterpiece of Greek architecture that was built in the fifth century B.C. The surface of the historic monument has been deteriorating for decades. It's time for temple-saving action.

The workers know that they have a hard job ahead. Work on another famous Greek temple, the Parthenon, has been going on for nearly 60 years, and it's not done yet.

These buildings, like many monuments, are built of marble—one of the hardest stones. Why are they in need of restoration?

Wind and rain have always had an effect on buildings, but the main cause of deterioration is pollution. The problem is not just in Athens. In cities around the world, historic buildings are literally being dissolved away.

The major culprits are acid rain and smog (visible as a reddish brown haze), which is a problem in most of the world's large cities. Both originate with the burning of fossil fuels, such as coal and petroleum. As these fuels burn, they give off gases, which include the pollutants sulfur dioxide and nitrogen oxides (nitrogen oxide and nitrogen dioxide).

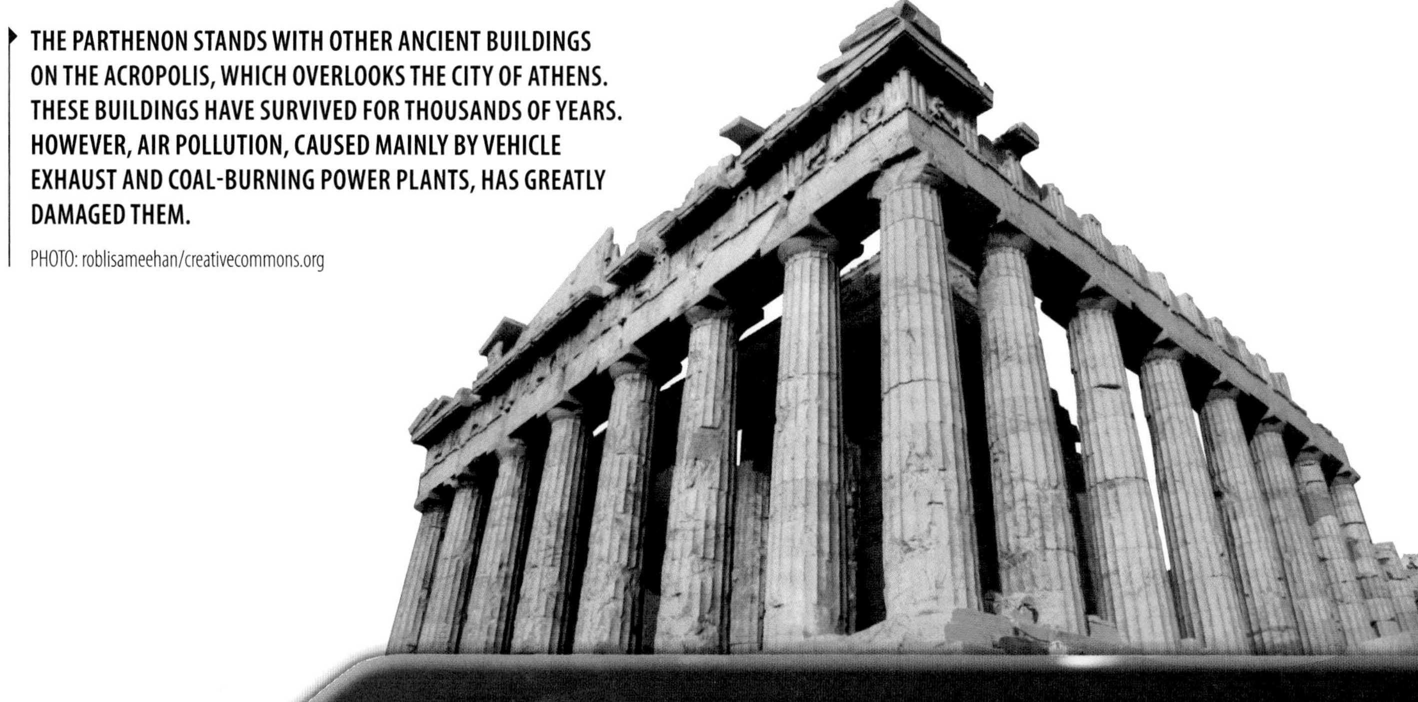

THE PARTHENON STANDS WITH OTHER ANCIENT BUILDINGS ON THE ACROPOLIS, WHICH OVERLOOKS THE CITY OF ATHENS. THESE BUILDINGS HAVE SURVIVED FOR THOUSANDS OF YEARS. HOWEVER, AIR POLLUTION, CAUSED MAINLY BY VEHICLE EXHAUST AND COAL-BURNING POWER PLANTS, HAS GREATLY DAMAGED THEM.

PHOTO: roblisameehan/creativecommons.org

READING SELECTION

EXTENDING YOUR KNOWLEDGE

ACID RAIN HAS DISSOLVED PARTS OF THIS STATUE.

PHOTO: Nino Barbieri/creativecommons.org

Sulfur dioxide is produced in particularly large quantities by coal-burning power plants and other industrial operations. Coal contains some sulfur and, as it burns, the sulfur reacts with oxygen from the air to form sulfur dioxide.

While petroleum does not contain nitrogen itself, the nitrogen and oxygen in the air react at the high temperatures that occur from the burning of gasoline in automobile engines. One major source of nitrogen oxides therefore is auto exhaust fumes.

These gases rise into the atmosphere where they combine with oxygen and water vapor. The sulfur dioxide becomes sulfuric acid, and the nitrogen oxides become nitric acid. Together, they form an acid solution that falls to earth as acid rain (or acid sleet or snow).

All rain is slightly acidic, but acid rain does much more damage to buildings. It is especially harmful to buildings made from rocks that contain calcium carbonate or magnesium carbonate. Marble, used in many Athenian buildings, and the softer, even more vulnerable limestone both contain carbonates. As years pass, the acid solution reacts with the surfaces of monuments and buildings and turns them into soluble substances. Acid rain can also attack paint and metals, and it forms a crust on the surface of glass.

SOME OF THE DAMAGE CAUSED TO THE PARTHENON, SUCH AS THE DISCOLORATION OF THE MARBLE, IS THE RESULT OF THE ACTION OF ACID RAIN DISSOLVING THE MARBLE FROM WHICH IT WAS BUILT.

PHOTO: Jeremy Rover/creativecommons.org

WHAT CAN YOU DO ABOUT ACID RAIN?

- Use the car less. Use public transportation, bike, walk, or carpool.
- Conserve electricity. Although alternative energy production from renewable sources such as wind has been gaining a foothold, much of our electricity is still produced by coal-burning power plants.
- Study historical sites, buildings, or cemetery headstones in your area. Discuss with your local historical society how you can help protect them.
- Contact a local environmental group to see whether it is taking action against acid rain and how you might help.
- Contact your local government (civic association, Mayor's office, Board of Supervisors, City Council, etc.) to see what actions have been taken in your community to reduce emissions of polluting gases.
- Discuss other actions you and your classmates can take to help solve this environmental issue.

READING SELECTION

EXTENDING YOUR KNOWLEDGE

DESPITE IMPROVED REGULATION OF EMISSIONS, MOTOR VEHICLES ARE A MAJOR SOURCE OF THE AIR POLLUTION THAT CAUSES ACID RAIN AND CONTRIBUTES TO GLOBAL WARMING. AIR POLLUTION IN SANTIAGO, CHILE, CONTRIBUTES TO DENSE SMOG THAT ON SOME DAYS OBSCURES THE VIEW OF THE ANDES MOUNTAINS.

PHOTO (right): Simone Ramella/creativecommons.org
PHOTO (below): Phil Whitehouse/creativecommons.org

THIS INDUSTRIAL SITE IS BELCHING OUT SMOKE AND GASES, INCLUDING SOME THAT CAUSE ACID RAIN. THOSE POLLUTANTS ARE RESTRICTED IN THE UNITED STATES, BUT ARE STILL COMMON IN SOME OTHER COUNTRIES IN THE WORLD.

PHOTO: Richard Costello/creativecommons.org

Not only does acid rain harm buildings, it damages trees and kills aquatic life and other organisms. To fight these effects, people around the world are applying a great deal of ingenuity to solve the problem of acid rain. In many countries, fossil-fuel-burning power plants and other industrial plants now remove some acidic gases from the waste products that would otherwise be dispersed through smokestacks. Also, special devices are being fitted to car tailpipes to remove some of these gases from exhaust fumes.

Until the source of the pollution is completely removed, any efforts to restore ancient buildings will be only stopgap measures. The team of workers on the Acropolis in Athens, in other words, is dealing with the symptoms, but not the cure. ■

DISCUSSION QUESTIONS

1. What chemical reactions are responsible for damage to old stone buildings?
2. In 1995, the Environmental Protection Agency (EPA) began an Acid Rain Program designed to reduce the emissions of sulfur dioxide and nitrogen oxides. Use Internet resources to find out whether the benefits of the program outweigh its costs.

LESSON 11

HOW MUCH SOLUTE DISSOLVES IN A SOLVENT?

SOLID SOLUBLE SALTS AND WATER ARE SEEN NEXT TO EACH OTHER IN THIS LAKE IN CALIFORNIA. HOW CAN SOLID SOLUBLE SALTS AND WATER EXIST IN THE SAME PLACE?

PHOTO: the_lazy_daisy/creativecommons.org

INTRODUCTION

As you know from Lesson 10, solutions are made from solvents and solutes. When you add a spoonful of common salt (sodium chloride) to a pan of water, it dissolves. Salt is soluble in water. Add a second spoonful, and that also dissolves. But what would happen if you kept adding salt? Would it continue to dissolve? Could you add more salt than there was water, or would the salt eventually stop dissolving? What would happen if you used a soluble substance other than salt? Would the same amount of that substance dissolve? These are some of the questions you will try to answer in this lesson. You will start by examining a blue liquid and explaining your observations of the liquid based on what you already know. You will then investigate two white crystalline substances. One is sodium chloride, and the other is sodium nitrate. They look almost the same, but as you will discover, they have different characteristics when they are added to water. Could these different characteristics be used to help identify these two substances?

OBJECTIVES FOR THIS LESSON

- Make solutions using different amounts of solute.
- Discover what is meant by the term "saturated solution."
- Design and conduct an experiment with your class to determine the solubility of two different substances.
- Discuss the design of your inquiry.
- Discuss solubility as a characteristic property of matter.

MATERIALS FOR LESSON 11

For you

1 copy of Student Sheet 11.1: Saturating a Solution
1 copy of Student Sheet 11.2: Determining Solubility
1 pair of safety goggles

For you and your lab partner

1 100-mL graduated cylinder
2 test tubes
1 test tube rack
2 rubber stoppers
Access to an electronic balance
Access to water

For your group

1 test tube containing a blue liquid
1 lab scoop
1 jar containing sodium chloride
1 jar containing sodium nitrate

GETTING STARTED

1. One student from your group should obtain the plastic box containing the materials.

2. Take out the test tube rack and the test tube containing the blue liquid. Pass the test tube around your group so that each member of your group can examine it closely. Discuss with your group precisely what you observe in the tube. Write your observations in your science notebook. What can you conclude from your observations?

3. Participate in a class discussion of your observations.

4. Before proceeding with Inquiry 11.1, hand in the test tube containing the blue liquid. Return the test tube rack to the plastic box.

SAFETY TIP

Wear your safety goggles at all times.

HOW MUCH SALT CAN YOU GET TO DISSOLVE IN A CUP OF WATER?

PHOTO: Dubravko Sorić/ creativecommons.org

SATURATING A SOLUTION

PROCEDURE

1. Check the materials in your plastic box against the materials list. You and your partner should have two test tubes, two rubber stoppers, a test tube rack, and a graduated cylinder. You will share the lab scoop and the jars of sodium chloride and sodium nitrate with the other pair in your group.

2. How much salt (sodium chloride) can you get to dissolve in a test tube filled halfway with water? Fill one test tube halfway with water. Add one level lab scoop of salt to the test tube. Place a rubber stopper in the test tube, and shake the mixture to help the salt dissolve faster. If it completely dissolves, add more salt. Keep adding salt until no more dissolves.

3. Answer the following questions on Student Sheet 11.1: Saturating a Solution:

   - How many scoops of sodium chloride dissolved in the water?
   - How did you know that no more would dissolve?

4. After a short class discussion, record your definition of a saturated solution.

5. Think about how you could adapt the technique you used in Step 2 to find out how many grams of sodium chloride dissolved in water.

6. Rinse the test tube with water. Put the test tube in the test tube rack.

INQUIRY 11.2

DETERMINING SOLUBILITY

PROCEDURE

1. Using the apparatus you have been given, how could you compare how much of each of the two substances (sodium nitrate and sodium chloride) will dissolve in water? Here are some questions you need to discuss with your partner. Write your answers in your science notebook.

   A. What will you need to measure?

   B. How will you know when you have a saturated solution?

   C. How will you calculate the amount dissolved?

2. Your teacher will conduct a short brainstorming session. Be prepared to contribute to the discussion. By the end of the brainstorming session, the class will have agreed on a procedure for determining solubility.

3. Answer the questions in Steps 1 through 3 on Student Sheet 11.2: Determining Solubility:

   - What are you trying to find out?
   - What materials will you use?
   - What is your procedure?

4. Under Step 4 of Student Sheet 11.2, design a data table to record your results and calculations.

5. Follow the class procedure for determining solubility, and record your results in the data table. When you have finished, pour the solutions down the drain with lots of water. Clean the test tubes and return the materials to the plastic box.

6. Under Step 5 on Student Sheet 11.2, calculate the number of grams of each substance that dissolved in the water and answer the following question: Are the different substances equally soluble in water?

7. Under Step 6 on Student Sheet 11.2, list any problems you had with the experiment or the approach. Could any of these problems have affected your results? Explain how.

READING SELECTION

BUILDING YOUR UNDERSTANDING

SOLUBILITY AND SATURATED SOLUTIONS

At room temperature, a solvent (such as water) can dissolve only a certain amount of solute. For example, in Inquiry 11.1, after adding a few lab scoops of sodium chloride to the water, you could see a white solid (undissolved sodium chloride) at the bottom of the tube. The white solid indicated that the water could not dissolve any more sodium chloride. When this happens, the solution is "saturated." The mass of solute dissolved in a given volume or mass of a solvent in a saturated solution is the solubility of the solute in that solvent. Solubility is usually measured in grams of solute per unit volume of solvent (for example, grams per liter) or in grams per 110 g of solvent. The solute that does not dissolve, or go into solution, is known as the "precipitate." When have you heard the term "precipitation" used? Do you see how it is related to the chemical term for something that falls to the bottom?

The solubility of a solute changes with changing temperature. For example, sodium nitrate becomes more soluble as the temperature rises. It is about twice as soluble at 80°C as it is at 1°C. There are some substances that become less soluble as the temperature rises. When you heated water in Lesson 7, you may have noticed that bubbles appeared, even though the water was well below the boiling point. These were bubbles of gases, such as oxygen and nitrogen, that were dissolved in the water. The gases became less soluble as the water was heated, and they were released from solution. ■

REFLECTING
ON WHAT
YOU'VE DONE

1. You will have an opportunity to look at the results of other pairs. Be prepared to discuss how these results could give a more accurate measure of the solubility of these two substances.

2. Answer the following question on Student Sheet 11.2: How could you use the property of solubility to help you identify a type of matter?

3. Read "Solubility and Saturated Solutions."

LESSON 12

MASS, VOLUME, AND DISSOLVING

WHAT HAPPENS TO THE MASS OF A SOLUTE WHEN THE SOLUTE IS ADDED TO A SOLVENT?

PHOTO: National Science Resources Center

INTRODUCTION

What happens to the mass and volume of two substances when the substances are mixed to form a solution? Will the mass and volume of the solute and the solvent remain the same before and after dissolving? In this lesson, you will conduct two inquiries. In the first inquiry, you will make a solution from two liquids of known mass and volume and compare their masses and volumes before and after mixing. In the second inquiry, you and your lab partner will devise a procedure for determining whether any change in mass occurs when salt (sodium chloride) is dissolved in water.

OBJECTIVES FOR THIS LESSON

- Predict what happens to the mass and volume of a solute and a solvent when these substances are mixed together to form a solution.
- Perform an inquiry to test your predictions.
- Design and conduct an inquiry to investigate whether a change in mass occurs when sodium chloride dissolves in water.

MATERIALS FOR LESSON 12

For you

1	copy of Student Sheet 12.1: Mixing Water and Alcohol
1	copy of Student Sheet 12.2: Dissolving a Solid and Measuring Mass
1	pair of safety goggles

For your group

2	250-mL beakers
4	100-mL graduated cylinders
2	pipettes
4	test tubes
1	lab scoop
1	jar containing sodium chloride
1	bottle containing ethyl alcohol (ethanol)
4	paper towels
	Access to water
	Access to an electronic balance

GETTING STARTED

1. Previously in the unit, you learned that two properties of matter are mass and volume. Why is mass, and not volume, used to measure the amount of matter in an object? Discuss this question with your group. You will be expected to contribute your ideas to a class discussion about mass and volume.

2. One member of your group should obtain the plastic box containing the materials. Check the contents of the plastic box against the materials list.

3. You will be working in pairs. Your group will share the lab scoop, the bottle of ethyl alcohol, and the jar of sodium chloride. Split the remaining apparatus equally between the pairs in your group.

4. Fill the beaker with water. You will use this water to practice pouring an exact volume of water into a 100-mL graduated cylinder.

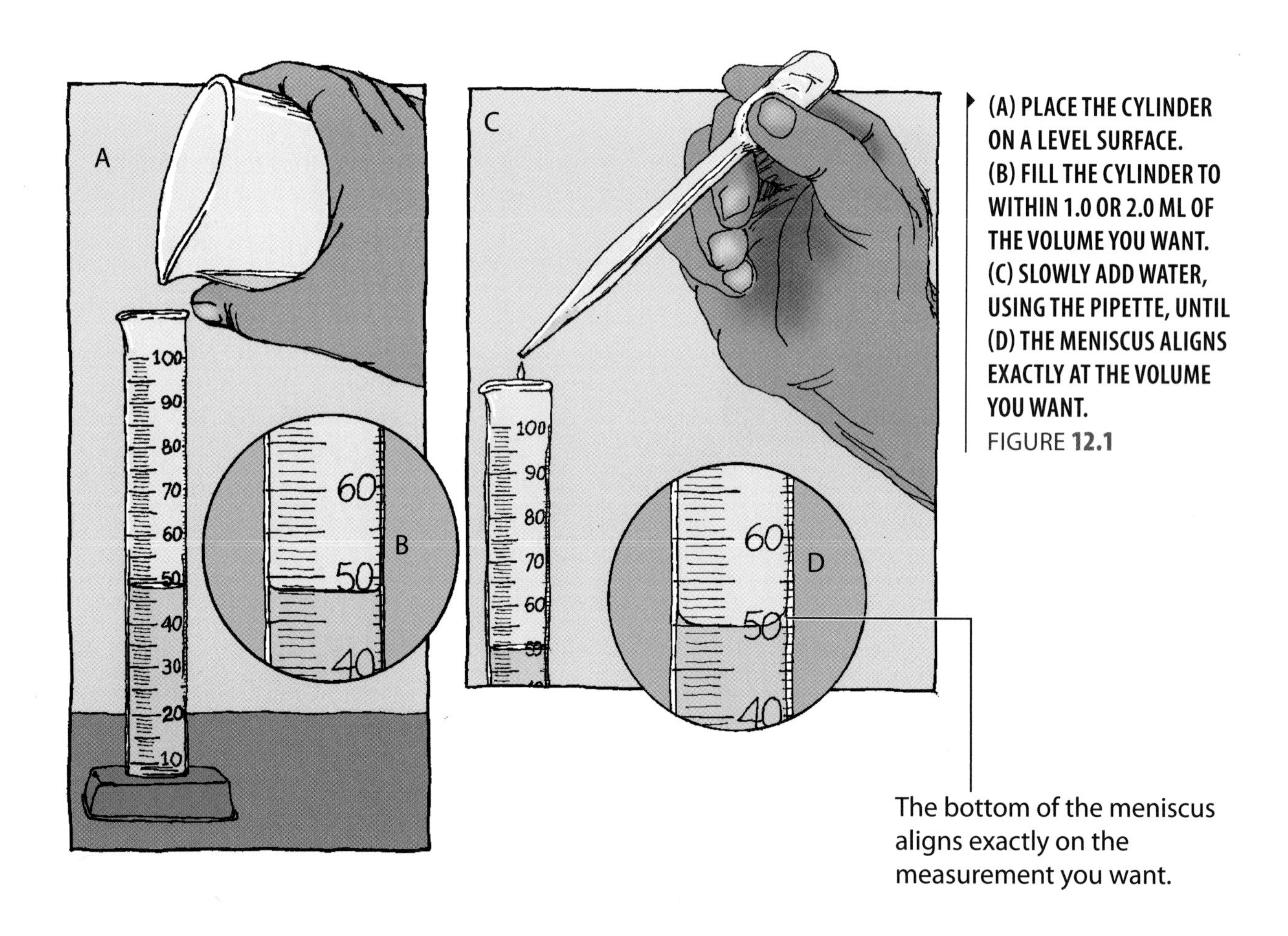

(A) PLACE THE CYLINDER ON A LEVEL SURFACE.
(B) FILL THE CYLINDER TO WITHIN 1.0 OR 2.0 ML OF THE VOLUME YOU WANT.
(C) SLOWLY ADD WATER, USING THE PIPETTE, UNTIL
(D) THE MENISCUS ALIGNS EXACTLY AT THE VOLUME YOU WANT.
FIGURE **12.1**

5. With your partner, review and practice the correct technique (refer to Figure 2.2 in Lesson 2 and Figure 12.1 in this lesson to ensure an accurate measurement), as follows:

   A. One partner chooses an exact volume of water for the other partner to pour into the graduated cylinder.

   B. Place the graduated cylinder on a level surface.

   C. Fill the graduated cylinder to 1.0 or 2.0 mL below the volume you require.

   D. Using the pipette, slowly add water until you have the exact volume you want.

6. Carefully check your partner's measurement.

SAFETY TIP

Wear your safety goggles throughout both inquiries.

INQUIRY 12.1

MIXING WATER AND ALCOHOL

PROCEDURE

1. Put exactly 50.0 mL of water into one of the 100-mL graduated cylinders.

2. Put exactly 50.0 mL of ethyl alcohol into the other 100-mL graduated cylinder.

SAFETY TIP

If you spill the ethyl alcohol, immediately tell your teacher.

3. Measure the mass of each cylinder and its contents. Record your results in Table 1 on Student Sheet 12.1: Mixing Water and Alcohol.

4. Predict what you think the volume will be after you mix the water and the ethyl alcohol. Predict what you think the mass will be after you mix the water and the ethyl alcohol.

5. Record your predictions in Table 1 on Student Sheet 12.1.

Inquiry 12.1 continued

6 Test your predictions by carefully pouring the ethyl alcohol into the 100-mL cylinder containing the water (see Figure 12.2). Allow a minute for the ethyl alcohol to drain completely from the graduated cylinder. Gently tap the cylinder with your finger to speed up the process. Take care to avoid spills.

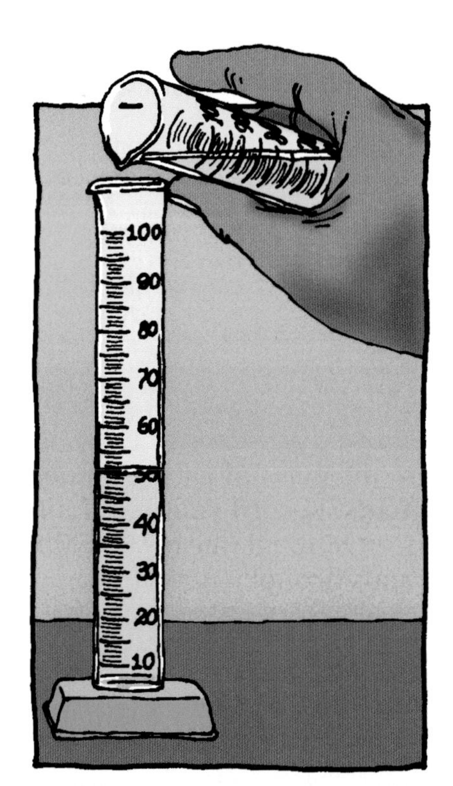

CAREFULLY MIX THE TWO LIQUIDS. ALLOW A MINUTE FOR THE ETHYL ALCOHOL TO DRAIN COMPLETELY FROM THE CYLINDER. GENTLY TAP THE CYLINDER WITH YOUR FINGER TO SPEED UP THE PROCESS. TAKE CARE TO AVOID SPILLS.
FIGURE **12.2**

7 Measure the volume of the mixture. Measure the combined mass of the graduated cylinder with the mixture and the empty graduated cylinder. In Table 1, record your measurements and calculate any differences in mass and volume before and after mixing. Empty the graduated cylinder into the appropriate container.

8 Answer the following questions in Steps 2a-2c on Student Sheet 12.1:

A. What type of mixture was formed when you mixed the water and the ethyl alcohol?

B. What happened to the volume?

C. What happened to the mass?

9 Write your results in the class results table (on the board or transparency).

10 Compare your results with those of the rest of the class. What conclusions can you reach? Record your conclusions. Be prepared to explain them during a class discussion.

11 Clean up according to the procedure specified by your teacher.

INQUIRY 12.2

DISSOLVING A SOLID AND MEASURING MASS

PROCEDURE

1. You and your lab partner will design an inquiry to determine what happens to the mass of sodium chloride and water when sodium chloride is dissolved in water. The following questions may help you in the design process:

   A. What do you need to measure?

   B. What apparatus should you use? (You may use any of the materials that are in the plastic box.)

   C. How much solute and solvent will you use? (Remember to consider the solubility of the sodium chloride.)

   D. What precautions should you take to ensure accurate measurements?

   E. How will you record your results?

2. Write your procedure on Student Sheet 12.2: Dissolving a Solid and Measuring Mass. If you have any problems, discuss them with your teacher.

3. Conduct the inquiry.

4. Record your results on Student Sheet 12.2. Compare them with those obtained by other pairs.

5. Clean and dry your apparatus. Return it to the plastic box.

6. What can you conclude from this experiment? Record your conclusions.

7. Participate in a class discussion about the procedure you used, your results, and your conclusions.

REFLECTING ON WHAT YOU'VE DONE

1. Answer these questions in your science notebook:

   A. What have you discovered by doing these two inquiries?

   B. What happens to the mass of two types of matter when they are mixed together to form a solution (for example, when a solid is dissolved in a liquid)?

   C. Does the same rule apply to volume?

2. How do your results compare with what you already know about what happens to the mass and volume of matter during phase change (Lesson 8)? Participate in a class discussion.

LESSON 13

RESEARCHING SOLVENTS

INTRODUCTION

How long does a pair of shoes last? How many years does it take for a carpet to wear out? Which toothbrush should I buy? These are typical questions that consumers ask. But where would you look to find the answers? Well, you could look in a consumer magazine. These magazines list a range of products—such as cameras, computers, toothbrushes, and household cleaners—and then score them according to how well they work.

Where do the magazines get their information? One approach is to collect details on consumers' experiences with a product. The magazine staff compiles the information to produce a rating. (For example, "Seven out of 10 consumers scored the Zippoteeth electric toothbrush the best, while the Scrubboplaque scored the lowest.")

Another, more scientific, approach is to test the products. Each product is put through a series of tests carefully designed to determine how well it does its job. Each product is then given a score using a predetermined scoring system. Results for the electric toothbrushes might look like this: "Zippoteeth scored 5 on plaque removal and 10 on battery life. Scrubboplaque scored 6 on plaque removal but only 2 on battery life."

Do you think designing the tests that go into making this sort of report is easy? Scientists must take into account many factors and must standardize the tests so they are fair to all the products being tested. Could you design tests like this? In today's lesson, you will have the chance to solve a similar type of problem. This one involves stain removal as well as some topics you investigated in previous lessons—solutions and dissolving.

THIS SCIENTIST IS TESTING BACTERIA TO SEE HOW WELL THEY WORK AT STOPPING BEETLE PESTS FROM EATING POTATO CROPS. HOW DO SCIENTISTS DESIGN TESTS SUCH AS THIS ONE TO MAKE SURE THEY ARE FAIR?

PHOTO: Peggy Greb, Agricultural Research Service/U.S. Department of Agriculture

OBJECTIVES FOR THIS LESSON

- Discuss solvents and their uses.
- Design and conduct an inquiry on stain removal.
- Present your results to the rest of the class.

MATERIALS FOR LESSON 13

For you

1 pair of safety goggles

For your group

3 dropper bottles containing these solvents:
Water
Rubbing (isopropyl) alcohol
Kerosene

3 plastic cups containing these staining substances:
Ketchup
Chocolate syrup
Vegetable oil

1 black permanent marker
1 ballpoint pen
5 white cotton cloth squares
10 cotton swabs
2 sheets of newspaper
1 sheet of newsprint
Masking tape

GETTING STARTED

1. Think of examples of liquids that do not contain water. Write your examples in your science notebook. You will be asked to contribute your ideas during a short brainstorming session.

2. After the brainstorming session and discussion, copy the diagram the class has produced into your notebook.

SAFETY TIPS

Wear your safety goggles throughout the inquiry.

Take care not to get solvents or stains on your clothes.

Do not taste any of the substances.

If you spill rubbing alcohol or kerosene, immediately tell your teacher.

THIS SCIENTIST KNOWS THAT PRODUCT TESTING REQUIRES THAT MANY FACTORS ARE TAKEN INTO ACCOUNT, AND THAT STANDARDIZED TESTS SHOULD BE USED.

FIGURE **13.1**

REMOVING STAINS

PROCEDURE

1. One member of your group should obtain a plastic box of materials. Check the contents of the box against the materials list.

2. Using the materials in the plastic box, your group should design a test that can be used to compare the effectiveness of the three solvents (in the bottles) at removing five different types of stain. The stains are ketchup, chocolate syrup, vegetable oil, permanent marker ink, and ballpoint pen ink. Your teacher may give you some other stains as well. Think about the following questions and test design considerations and discuss them with your group:

   A. How will you standardize your testing procedures so that the results obtained for each solvent and stain can be fairly compared? What elements will you need to standardize?

   B. The stains will need to be dry before you test them. How will you accomplish this?

   C. How will you score the effectiveness of the stain removers on each stain?

   D. How will you present your results so they are easy for others to understand?

   E. How will you divide the work among the members of your group?

   F. How long will you take to conduct each step of the procedure? (Your teacher will tell you how much total time you have.)

Inquiry 13.1 continued

3. In your science notebook, write what you are trying to find out. Agree on the materials you are going to use, a procedure, the design of a scoring rubric (system), and a results table. Record this information under the following headings: Materials, Procedure, Scoring Rubric, and Results Table.

4. Draw a large version of your group's results table on the sheet of newsprint.

5. Apply the stains. Write the names of the members of your group on the cloth squares. Allow the stains to dry in the place suggested by your teacher.

6. Continue with your procedure during the next period if necessary.

7. Transfer all of your results to the table on the newsprint. Tack or tape the newsprint on the wall nearest your table. Make sure you also copy all the results into your science notebook.

8. Clean up the materials. Dispose of the cotton swabs, cloths, and small containers of staining substances. Return the remaining items to the plastic box.

REFLECTING ON WHAT YOU'VE DONE

1. Discuss the results with your group. In your science notebook, write any conclusions you can make from your test data. Include any comments or suggestions about the effectiveness of your procedure.

2. Your teacher will lead a class discussion. One member of your group will be asked to report on your procedure, scoring rubric, results, and conclusions.

Getting Taken to the Cleaners

Have you ever bought a new piece of clothing, worn it once, put it through the wash, and when you've tried to wear it again, discovered it had shrunk or was completely misshapen? It's only then that you bother to read the label—"Dry clean only."

Dry cleaning is used to clean clothes that would be harmed by water. It is also used to remove stains that are insoluble in water (for example, grease). As the name suggests, dry cleaning involves cleaning without water. (Actually, a very small amount of water is used—you'll find out why later—but not enough to change the name to wet cleaning!) However, even though only a little water is used, the term "dry cleaning" is still a bit deceptive. That's because liquids other than water are used.

Early forms of dry cleaning used petroleum solvents such as kerosene. But kerosene is flammable—it burns. After a series of explosions at dry cleaners, the solvent tetrachloroethylene was widely adopted, and it is still used today (along with other solvents). Tetrachloroethylene is not flammable, but its fumes can be toxic in enclosed spaces. That is one reason why, if dry-cleaned clothes smell strongly of solvent, you should drive home from the dry cleaners with your car windows open.

How does dry cleaning work? A dry-cleaning machine is like a giant washing machine. Clothes are placed in the machine. Tetrachloroethylene, mixed with a very small amount of water

READING SELECTION

EXTENDING YOUR KNOWLEDGE

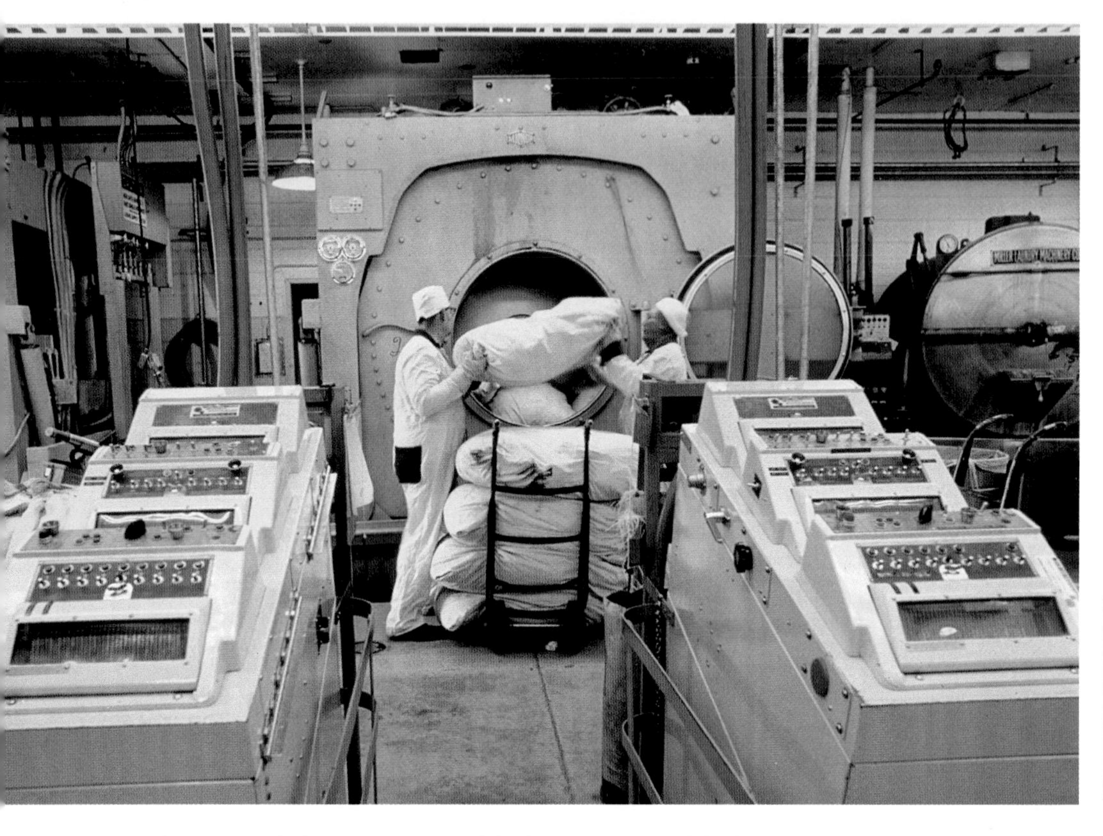

DRY-CLEANING MACHINES WORK SOMEWHAT LIKE WASHING MACHINES, BUT THEY USE A SOLVENT OTHER THAN WATER.

PHOTO: U.S. Department of Energy

and a special detergent, is added. (Water is added to remove any stains caused by water-soluble substances.) The tetrachloroethylene flows continuously through the machine until the clothes are clean. Any solvent that remains in the clothes after the cleaning cycle eventually evaporates. All the remaining solvent is recycled using distillation to clean it. After every load, it is heated until it evaporates, leaving behind the particles of dirt, and then cooled until it condenses to produce a clean solvent that can be used again.

However, a new approach to dry cleaning has been developed. This method does not use any toxic dry-cleaning solvents. Special detergents and carbon dioxide, which is the solvent, clean the clothes. The carbon dioxide, a gas normally found in air, is put under pressure during the cleaning process. This pressure keeps the carbon dioxide in a liquid state. Both the special detergents and the carbon dioxide can be recycled. Will this more environmentally friendly approach be the future of dry cleaning? Based on what you know about global climate change, why do you think it is important to recycle the carbon dioxide? ■

DISCUSSION QUESTIONS

1. Why do people who own washing machines still go to the dry cleaners?
2. What are the advantages of recycling the solvents used in the dry-cleaning process?

Mixing Colorful Coverings

WHY DO WE USE PAINT AND WHAT DOES IT CONSIST OF?

PHOTO: The Marmot/creativecommons.org

Wouldn't the world be a dull place if there were no such thing as paint? Since prehistoric times, paints have been used in art. The earliest cave paintings were made using paints made from colored soils and rocks or from animals and plants. These were mixed with other substances, such as egg whites or spit, which allowed the pigments to spread over and stick to the surface being painted. Today, paints are used to protect and decorate surfaces. They are carefully formulated mixtures, designed to do specific jobs, and are available in a seemingly infinite variety of colors. Let's examine these mixtures more closely and see how the different properties of the substances from which they are made work together.

READING SELECTION

EXTENDING YOUR KNOWLEDGE

THROUGHOUT HISTORY, ARTISTS HAVE USED A WIDE VARIETY OF PIGMENTS IN THEIR PAINTINGS. MANY PAINTS CONTAIN OXIDES OF METALS, WHICH PROVIDE COLOR. BUSHMAN ROCK ARTISTS USED SOILS CONTAINING REDDISH BROWN IRON OXIDE TO PAINT THESE IMAGES. MANY MODERN OIL PAINTS CONTAIN OTHER MATERIAL OXIDES THAT PRODUCE THE VIVID COLORS ASSOCIATED WITH OIL PAINTINGS.

PHOTO: © David Marsland

THIS ARTIST IS PAINTING A MURAL. ARTISTS' PAINTS CONTAIN MANY DIFFERENT SOLVENTS. WATER AND TURPENTINE ARE THE MOST COMMON.

PHOTO: National Archives and Records Administration

Most paints consist of pigments, a vehicle, and a solvent, plus other additives that perform a variety of functions. Pigments give the paint color and also make it opaque (not transparent). The type of pigment used depends on the color of the paint desired. For example, white paint often contains the pigment titanium dioxide. However, several pigments can be used together in varying quantities to produce a wide range of paint colors. For instance, even though titanium dioxide is used to make white paint, other pigments are often mixed with it to produce paints of other colors. For example, titanium dioxide is mixed with barium chromate or cadmium sulfide to make yellow paints, with chromium oxide to make green paints, and with ultramarine or dyes such as indanthrene blue to make blue paints. In addition to adding color to paint, titanium dioxide has the ability to hide the surface that is being painted. For this reason it is called a hiding pigment.

Paint must also contain a substance that will make the pigment stick to the surface being painted (like the egg whites and spit used by prehistoric cave painters). These bonding substances are called the vehicle. They are usually made from plastic-like substances which, when dry, form a hard, flexible protective coating.

Solvent thins the paint and helps it spread during painting. Mineral spirits (a distilled fraction of petroleum) are used as solvents in some glossy paints. These dissolve the vehicle. When the paint dries, the mineral spirits evaporate, and a hard film is left behind.

Many emulsion paints and modern latex paints use water as a thinner, although in these cases, the water does not dissolve the vehicle but keeps it finely divided. When the paint dries, the water evaporates and the solid particles come together to form a hard, flexible surface.

Paints often contain a variety of additives that perform various functions. They may improve the weather resistance of the paint, affect the way the pigment is dispersed to produce special finishes, or speed up the drying process. ■

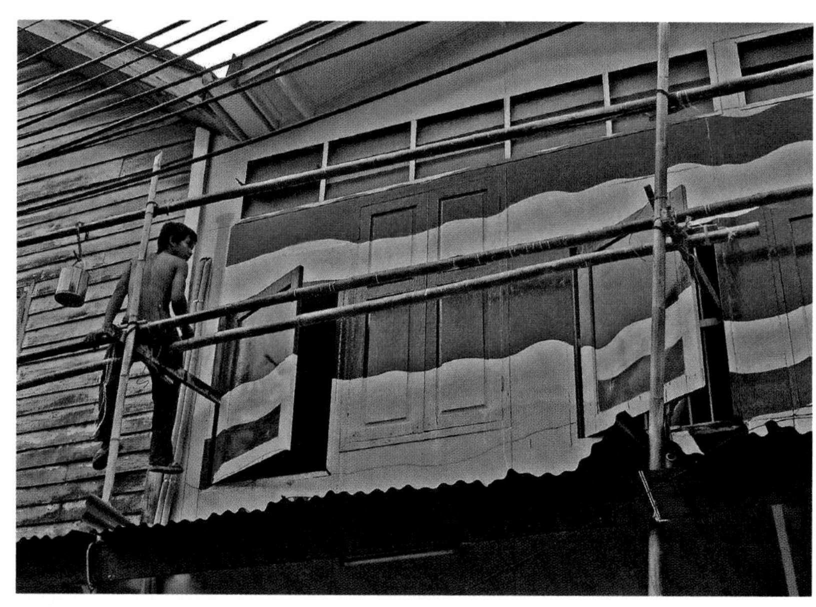

PAINT IS A COMPLEX MIXTURE DESIGNED TO BE APPLIED AS A LIQUID AND TO DRY INTO A DECORATIVE AND HARD, BUT FLEXIBLE, PROTECTIVE FINISH.

PHOTO: Michael McCauslin/creativecommons.org

DISCUSSION QUESTIONS

1. Spray paint is convenient for small jobs. Research whether spray paint contains the same basic ingredients as other paints.
2. What are some mixtures (besides paint) that we rely on for maintaining our homes?

LESSON 14

END-OF-UNIT ASSESSMENT

INTRODUCTION

This lesson is the assessment for the unit *Exploring the Properties of Matter*. The assessment has two sections. In Section A, you will work by yourself to investigate a mystery object. You will use your measurement skills, your knowledge of density, and a data table to determine the substance that makes up your mystery object. Section B consists of several multiple-choice questions. Some of these will require you to use your knowledge and skills to interpret data tables, diagrams, graphs, and experiments. Your teacher will use the results of this assessment to evaluate how well you can apply the concepts, knowledge, and skills you learned in the unit.

CAREFUL LABORATORY WORK IS AN IMPORTANT PART OF SCIENCE. IN THIS LESSON, YOU WILL TEST SOME OF YOUR LAB SKILLS.

PHOTO: Courtesy of DOE/NREL, Credit—Rick Matson

OBJECTIVES FOR THIS LESSON

- Discover the identity of the matter that makes up your mystery object.
- Use your knowledge and skills to solve problems related to the characteristic properties of matter.

MATERIALS FOR LESSON 14

For you

1	copy of Student Sheet 14: Assessment
1	100-mL graduated cylinder
1	metric ruler
1	mystery object

For the class

8	loupes (double-eye magnifiers)
16	250-mL beakers
	Access to water
	Access to an electronic balance

GETTING STARTED

Your teacher will assign you to one set of apparatus. You will share loupes and beakers with the rest of the class. Your teacher will tell you which stations to use when you need these items.

Follow along as your teacher reviews the guidelines for the assessment:

A. Immediately check your apparatus against the materials list.

B. You will work individually and should not talk to other students.

C. Answer all of the questions.

D. Do the performance component first.

E. As soon as you finish the performance assessment, begin the written assessment.

F. You have 15 minutes to complete the performance part of the assessment and the rest of the class period to complete the multiple-choice questions.

G. Three or four minutes before the end of the lesson, hand in Student Sheet 14: Assessment. Follow your teacher's instructions for cleaning up.

HOW WOULD YOU FIND OUT WHAT SUBSTANCES THESE OBJECTS ARE MADE FROM?

PHOTO: psyberartist/creativecommons.org

WHAT SUBSTANCE MAKES UP MY MYSTERY OBJECT?

PROCEDURE

1. Use any of the apparatus, a balance, water, and Table 14.1 to identify the substance that makes up your mystery object. Your object is made from one of the substances in Table 14.1. You may not need to use all of the apparatus.

2. You may refer to your science notebook, student sheets, and Student Guide while working on Inquiry 14.1, but do not discuss your work with other students.

3. Write your name on Student Sheet 14: Assessment.

4. Write the number of your object in the space provided in Step 1 on Student Sheet 1.4.

5. When you complete Section A of the assessment, begin Section B.

TABLE 14.1 APPROXIMATE DENSITIES OF MATERIALS AT ROOM TEMPERATURE

NAME OF SUBSTANCE	APPROXIMATE DENSITY (g/cm^3)
POLYVINYL CHLORIDE (PVC)	1.5
TITANIUM	4.1
ALUMINUM	2.7
SULFUR (RHOMBIC)	2.1
MERCURY	13.5
STEEL	7.9
LEAD	11.4
COPPER	8.9
NYLON	1.2
WATER	1.0

Glossary

boiling: The process by which a liquid turns into a gas at its boiling point.

boiling point: The temperature at which a liquid turns into a gas. Boiling points depend on air pressure. Boiling points of substances are usually given for standard air pressure (1 atmosphere).

Celsius (°C): A temperature scale with the melting point of ice at 0 degrees and the boiling point of water at 100 degrees. The divisions on the Celsius scale are the same as those on the Kelvin scale. See also *Kelvin.*

characteristic property: An attribute that can be used to help identify a substance. A characteristic property is not affected by the amount or shape of a substance.

composite: A material made from two or more substances: for example, resin and glass fibers make the composite fiberglass.

condensation: The process by which a gas turns into a liquid.

convection current: Movement of a gas, liquid, or plastic solid caused by variations in density that result from uneven heating of matter.

density: The mass of a substance that fills one unit of volume. It is usually measured in grams per cubic centimeter (g/cm^3).

dissolving: The process that takes place when a solvent is mixed with a solute to make a solution. For example, when sodium chloride is mixed with water, it dissolves to form a solution of sodium chloride in water.

evaporate: To change from a liquid to a gas at or below the boiling point.

expansion: The increase in the volume of matter that occurs when matter is heated.

Fahrenheit (° F): A temperature scale with the melting point of ice at 32 degrees and the boiling point of water at 212 degrees. See also *Celsius*; *Kelvin.*

freeze: The change in state in which a liquid turns into a solid.

gas: A state or phase of matter in which a substance has no definite shape or volume. Oxygen is an example of a gas.

gram: A metric unit used to measure mass. See also *density.*

immiscible: A term used to describe liquids that are unable to dissolve in one another. See also *miscible.*

Kelvin: A temperature scale with the lowest possible temperature at the zero point, which is called absolute zero. On the Kelvin scale, ice melts at 273 K. See also *Celsius.*

liquid: A state or phase of matter in which a substance has a definite volume but no definite shape. Liquids take the shape of the part of the container they occupy.

mass: A measure of the amount of matter in an object. Mass is measured in grams or kilograms.

material: The substance from which something is made.

matter: Substances that make up the universe. All matter has mass and volume.

melting: The phase change in which a solid turns into a liquid.

melting point: The temperature at which a solid turns into a liquid. The melting point of a substance is the same temperature as its freezing point. Melting points of substances are altered by changes in pressure and are usually given for standard air pressure (1 atmosphere).

miscible: A term used to describe liquids that are able to dissolve in one another. See also *immiscible.*

phase or state: Solids, liquids, and gases are the three phases or states of matter. For example, in a mixture of ice and water, ice is the solid phase and water is the liquid phase. Water in the gaseous phase is called water vapor or steam.

physical properties: All the characteristic properties of a substance except those that determine how it behaves in a chemical reaction.

precipitate: Solid that separates out of a solution.

respiration: A series of chemical reactions that take place in the cells of organisms during which energy used for life processes is released.

saturated solution: A solution that will not dissolve any more solute at a given temperature or pressure. For example, if copper (II) sulfate is added to a test tube of saturated solution of copper (II) sulfate, the crystals will remain undissolved at the bottom of the test tube.

solid: A phase or state of matter in which a substance has definite shape and volume.

solute: The substance that dissolves in a solvent. Solutes may be solids, liquids, or gases.

solvent: The substance in a solution that dissolves the solute.

temperature: The measurement of how hot an object is. Temperature is measured using a temperature scale. See also *Celsius; Fahrenheit; Kelvin.*

thermal decomposition: A chemical reaction in which a compound is decomposed by heating.

volume: The amount of space occupied by a sample of matter. Volume is measured in liters (L) and milliliters (mL) as well as in cubic centimeters (cm^3) and cubic meters (m^3).

weight: A measure of the force of gravity. Like all forces, weight is measured in newtons (N).

Index

Photo Credits

Front Cover
NPS Photo by Frank Balthis

Lessons
2 NOAA's National Weather Service Collection **14** Jamie L. Fumo **15** NASA/WMAP Science Team **16** NASA/WMAP Science Team **18** Tim Sheerman-Chase/creativecommons.org **29** Courtesy of the Smithsonian Institution Libraries, Dibner Library of the History of Science and Technology, Washington, D.C. **30** U.S. Coast Guard photo by Public Affairs Specialist 3rd Class Adam Baylor **32** Brendan Adkins/creativecommons.org **35** Courtesy of NOAA and the Russian Academy of Sciences **36 (top)** Library of Congress, Prints & Photographs Division, LC-USZ62-116257 **(bottom)** U.S. Coast Guard **37** U.S. Navy photo by Photographer's Mate 2nd Class Andrew M. Meyers **38** OAR/ERL/National Severe Storms Laboratory (NSSL) **43 (top)** ©2009 Carolina Biological Supply Company **(bottom left)** ©2009 Carolina Biological Supply Company **(bottom center)** ©2009 Carolina Biological Supply Company **(bottom right)** ©2009 Carolina Biological Supply Company **44** Library of Congress, Prints & Photographs Division, LC-USZ62-115014 **45** Jeff Sandquist/creativecommons.org **46 (left)** Smithsonian Institution, Carl C. Hansen, Neg. #91-17588 **(right)** NASA Headquarters—Greatest Images of NASA **47** NASA Headquarters—Greatest Images of NASA **48** FEMA/Andrea Booher **49** Criss!/creativecommons.org **52** National Science Resources Center **58 (left)** Cynthia M. Green, GLA Agricultural Electronics **(right)** Elektra Noelani Fisher/creativecommons.org **60 (top left)** Library of Congress, Prints & Photographs Division, LC-USZ62-7923 **(top right)** Library of Congress, Prints & Photographs Division, LC-USZ62-124161 **(bottom left)** Göran Henriksson, Department of Physics and Astronomy, Uppsala University **(bottom right)** Library of Congress, Prints & Photographs Division, LC-USZ62-64292 **63 (left)** Laura Rush/creativecommons.org **(right)** San Diego Air and Space Museum **65** NASA Goddard Space Flight Center Scientific Visualization Studio, Data provided by: Norman King (NASA/GSFC) **66** Rupert Taylor-Price/creativecommons.org **68 (top)** University of Colorado, National Geophysical Data Center, NOAA **(bottom)** IKONOS satellite imagery courtesy of GeoEye. Copyright 2008. All rights reserved. **69** OAR/National Undersea Research Program (NURP) **70** U.S. Air Force photo by Staff Sgt. Joshua Strang **71** Courtesy of Alyeska Pipeline Service Company **74** U.S. Navy photo by Mass Communication Specialist 3rd Class Paul J. Perkins **80** U.S. Air Force photo by Airman 1st Class Kathrine McDowell **81 (bottom left)** Andrew Silver, Mineral Collection of Brigham Young University Department of Geology, Provo, Utah/U.S. Geological Survey **(top right)** U.S. Geological Survey **82** LHOON/creativecommons.org **83 (bottom left)** Centers for Disease Control and Prevention/Dr. Edwin P. Ewing, Jr. **(top right)** Adrien Lamarre/U.S. Army Corps of Engineers **84** Library of Congress, Prints & Photographs

Division, LC-USZ62-72119 **87** ©2009 Carolina Biological Supply Company **88** ©2009 Carolina Biological Supply Company **90 (left)** Library of Congress, Prints & Photographs Division, LC-USZ62-54453 **(right)** National Oceanic and Atmospheric Association **91** Courtesy of Pennsylvania Historical & Museum Commission, Drake Well Museum, Titusville, PA **92** Courtesy DOE/NREL, Credit — David Parsons **93 (top left)** Eliot Elisofon, Photographic Archives, National Museum of African Art, Smithsonian Institution, EEPA 6963 **(bottom right)** Eliot Elisofon, Photographic Archives, National Museum of African Art, Smithsonian Institution, EEPA 6974 **95 (top)** Franko Khoury, National Museum of African Art, Smithsonian Institution, 85-19-12 **(bottom)** Investment Casting Institute **96** Randolph Femmer/National Biological Information Infrastructure **97** John M. Evans, U.S. Geological Survey, Colorado District **98** Courtesy of Lawrence Livermore National Laboratory **100** Commander John Bortniak, NOAA Corps **102** © David Marsland **107 (top)** Library of Congress, Prints & Photographs Division, LC-B2-943-5 **(bottom)** NASA Dryden Flight Research Center **108** U.S. Air Force Courtesy Photo **111** The Artifex/creativecommons.org **118 (top)** ©David Marsland **(bottom)** ©David Marsland **121** Courtesy of Cannondale, by Tim DeFrisco **124** Randolph Femmer/National Biological Information Infrastructure **126** S. Diddy/creativecommons.org **129** roblisameehan/creativecommons.org **130** Nino Barbieri/creativecommons.org **131** Jeremy Rover/creativecommons.org **132 (top right)** Simone Ramella/creativecommons.org **(bottom)** Phil Whitehouse/creativecommons.org **133** Richard Costello/creativecommons.org **134** the_lazy_daisy/creativecommons.org **136** Dubravko Sorić/creativecommons.org **140** National Science Resources Center **146** Peggy Greb, Agricultural Research Service/U.S. Department of Agriculture **152** U.S. Department of Energy **153** The Marmot/creativecommons.org **154 (top)** © David Marsland **(bottom)** National Archives and Records Administration **155** Michael McCauslin/creativecommons.org **156** Courtesy of DOE/NREL, Credit—Rick Matson **158** psyberartist/creativecommons.org